Validation of Computerized Analytical Systems

Ludwig Huber

Interpharm Press, Inc.
Buffalo Grove, IL

10 9 8 7 6 5 4 3 2 1

ISBN: 0-935184-75-9

Interpharm Press, Inc.
1358 Busch Parkway
Buffalo Grove, IL 60089, USA

Phone: + 1 + 708 + 459-8480
Fax: + 1 + 708 + 459-6644

Preface

Validation of computerized analytical systems is required by many regulations, quality standards and company policies. This book guides quality assurance and laboratory managers and users of computer-controlled analytical instruments through the entire validation process from design through implementation, testing and installation qualification to on-going calibration and performance qualification. For equipment purchased from a vendor it gives guidelines on the validation responsibilities of users and suppliers and on the qualification of vendors. It also gives guidelines on how to evaluate and validate existing computer systems retrospectively and how to validate "office software" used for laboratory applications such as spread sheets and databases and software that has been written by the user to further customize a system purchased from a vendor. It gives recommendations on how and how often to test and calibrate equipment in analytical laboratories and on how to validate analytical methods.

The book takes into account most national and international regulations and quality standards. Its concept and the examples and templates it includes are based on the author's multinational experience with all aspects of validation within Hewlett-Packard and from many personal discussions with regulatory agencies, equipment users, corporate QA managers, instrument vendors and consultants and seminars conducted by the author. Readers of the book will learn how to speed up their validation process, getting it right the first time, thereby avoiding troublesome rework, and gaining confidence for audits and inspections.

The book is divided in two parts: Part one deals mainly with validation of software at the vendor's and at the user's site. Part two covers validation and verification of complete systems, comprising software and measurement equipment. The focus of the book is more on the validation of software and computer part of the systems and not so much on measurement equipment. It is the author's experience that personnel in chemical laboratories are already more familiar with the verification, calibration and testing of measurement equipment than they are with the validation of software. Whereas the background information on software and computer system validation as well as the validation efforts are detailed, only general recommendations and strategies are given for the validation of measurement equipment.

The book uses mainly chromatographic instrumentation from Hewlett-Packard as examples. The strategies can be similarly applied to other analytical instruments and computer systems and to equipment from other vendors. The concepts and ideas expressed within this book are those of the author and do not necessarily reflect official Hewlett-Packard policy.

The book would not have its value without the generous contributions of experts on computer system and equipment validation. Their ideas and suggestions after proofreading draft versions have been implemented throughout the book. The author would like to thank Peter Bosshard of Hoffman La-Roche, Switzerland, for his inputs regarding the entire validation process of computer systems and to Rory Budihandojo of Warner Lambert, New Jersey and Alex Pavloff of Burroughs Wellcome, North Carolina, for their valuable inputs given from a pharmaceutical industry point of view. Naem Mady of Ciba Geigy, New Jersey, shared his experience with GLPs as practiced in the chemical industry.

Richard Johnson from the US Environmental Agency, NC, clarified aspects of EPA's Good Automation Laboratory Practices. Expert advice from an environmental testing laboratory point of view was readily forthcoming from M. Davies of Thames Water Utilities, U.K. Hans Biesel and Wolfgang Winter of Hewlett-Packard, Germany, shared their in-depth knowledge on software development, testing, documentation and support. Gordon Ross of Hewlett-Packard, Germany, contributed to the section on testing of capillary electrophoresis equipment.

The author would also like to express his thanks to the staff of Interpharm Press, in particular to Amy Davis, for help and kind support throughout the publication process. The work of the editorial manager, Jane Steinmann, is also greatly appreciated.

The author is indebted to the Hewlett-Packard company, Advanstar Communication, EURACHEM, the United States Environmental Protection Agency, and the UK Pharmaceutical Industry Computer System Validation Forum for their permission to make references to their documentation.

Ludwig Huber

Comments

This book is intended to help clarify some current issues in the area of validating computer-controlled analytical systems. Readers are encouraged to submit their comments and suggestions, especially if they have made different experiences in daily laboratory work. Comments should be submitted directly to the author.

Ludwig Huber
Hewlett-Packard Waldbronn Analytical Division
Postfach 1280
D-76337 Waldbronn
Germany

Fax: +49-7243-602501
Phone: +49-7243-602209

Contents

Preface iii

1. Introduction 1

2. Regulations and Quality Standards 11

(Current) Good Manufacturing Practice (cGMP) Regulations 12

Good Laboratory Practice (GLP) Regulations 16

Good Clinical Practice (GCP) Regulations 19

Good Automated Laboratory Practice (GALP) 20

Quality Standards and Guidelines 25

EN 45000 Series and ISO/IEC Guide 25 25

NAMAS Accreditation Standard 28

ISO 9000 Series of Quality Standards and ISO 9000-3 29

TickIT 29

3. Definitions: Computer Systems, Computerized Systems and Software 31

4. Validation Processes in the Analytical Laboratory 35

Definitions of Validation 35

Validation vs. Verification, Testing, Calibration and Qualification 38

Testing 38

Verification 39

Calibration 40

Qualification 41

Validation 41

Validation Steps in the Laboratory 42

5. **Life Cycle Approach for Validation of Software and Computer Systems** **45**

Validation of New Systems During Development 47

Evaluation of Existing Systems (Retrospective Validation) at the User's Site 48

Validation of 'Office Programs' Used for Laboratory Applications 54

Revalidation and Reverification of Software and Computerized Systems 56

When Is Reverification and Revalidation Required? 56

What Should Be Revalidated? 57

How Should Test Files Be Reused? 57

6. **Validation Efforts at the Vendor's Site** **59**

Requirements Analysis and Definition Phase 59

Design Phase 61

Implementation Phase 62

Test Phase 64

Types of Testing 64

Alpha-Testing 65

Beta-Testing 66

Defect Tracking and Response System 66

Release for Production and Installation 67

Operation and Maintenance 68

Change Control 69

Documentation 70

7. **Responsibilities of Vendors and Users** **71**

Software Categories 72

Validation Responsibilities 74

Source Code Availability 76

Vendor Testing and Specifications 77

User Testing 79

Support 79

Combined Standard and User Contributed Software 80

8. **Calibration, Verification and Validation of Equipment** 85

Validation/Verification at the Vendor's Site 87

Activities at the User's Site 89

9. **Selection and Qualification of a Vendor** 93

Selecting a Vendor 93

How Well Do They Know Good Practice Regulations and Quality Standards? 93
Will They Allow an Audit? 94
Do They Provide Adequate Long-Term Support? 94
Do They Build All the Required Functions into the Product? 94
Do They Supply Source Code to Regulatory Agencies? 94
Do They Supply Details of Algorithms? 94
Do They Have an Adequate Quality System? 95
Will They Asist with On-Site Validation? 95

Vendor Qualification 95

10. **Installation and Operation** 101

Preparing for Installation 101

Installation 102

Logbook 105

Operator Training 105

Preparing for Operation 107

Operation 107

11. **Validation of Analytical Methods** 115

Strategies for Method Validation 116

Parameters for Method Validation 116

Selectivity 118

Precision and Reproducibility 120
Accuracy 122
Linearity 122
Range 123
Limit of Detection and Quantitation 123
Stability 124
Ruggedness 124

12. Maintenance and Ongoing Performance Control 127

Maintenance 127
Calibration and Performance Verification 128
Calibration 128
Performance Verification (PV) 129
System Suitability Testing and Analytical Quality Control (AQC) 131
System Suitability Testing 131
Quality Control (QC) Samples with QC Charts 132
Handling of Defective Instruments 135

13. Testing of Chromatographic Computer Systems 139

Examples for Specifications and Tests for a Computerized HPLC System 141
Specifications 141
Modular Functional Testing 141
Integrated System Testing 143
Verification of Peak Integration 145
Automated Procedure for Testing a Chromatographic Computer System 147

14. Data Validation, Audit Trail, Security and Traceability 151

Data Entry 152
Manual Data Entry 153
Tracking Equipment and People 154

Raw Data: Definition, Processing and Archiving 154
Definition of Raw Data *154*
Defining and Archiving Chromatographic Raw Data *155*
Audit Trail for Amended Data *160*

Validation of Data 161

Security and Integrity of Data and Back-up 163
Security *163*
Back-up and Recovery *167*
Security Features Built into the Computer System *167*

15. Diagnostics, Error Recording and Reporting 171

16. Audit/Inspection of Computerized Systems 177

Appendix A. Glossary 185

Appendix B. (Standard) Operating Procedures 195

Example #1: Development and Validation of Simple Application Software 197

Example #2: Development and Validation of a Complex Application Software 200

Example #3: Testing Precision of Peak Retention Times and Areas of an HPLC System 204

Example #4: Retrospective Evaluation and Validation of Existing Computerized Analytical Systems 210

Appendix C. Strategy for Chronological Selection and Validation of Computerized Analytical Systems 217

Appendix D. Testing of Selected Equipment 221

High-Performance Liquid Chromatography 222
Precision of Retention Times and Peak Areas *223*
Baseline Noise of Systems with UV-visible Detectors *223*
Signal to Noise of Systems with Non UV-visible Detectors *224*

Limit of Detection 224
Limit of Quantitation 224
Linearity 224
Wavelength Accuracy of UV-visible Detectors 226
Carry Over 226
Flow Rate Accuracy 227
Composition Accuracy 227
Pressure Test 227

Capillary Electrophoresis (CE) 227
Migration Time Precision 228
Peak Area Precision 228
Carry Over 228
Detector Baseline Noise 228
Wavelength Accuracy 229
Detector Linearity 229
Injection Linearity 229

Gas Chromatography 229
Precision of Retention Times and Peak Areas 230
Signal to Noise of Detectors 230
Linearity 230
Oven Temperature Accuracy 230

References **233**

Indexes **245**

Name Index 245

Subject Index 247

1. Introduction

When Good Laboratory Practice (GLP) and Good Manufacturing Practice (GMP) regulations were first introduced, computers were not widely used in laboratories and so no special attention was paid to the use of computer hardware and software. They were treated like any other laboratory instrumentation and covered under regulations such as those GLPs in the US code of the federal register, for example, §58.61 and §58.63 on design, maintenance and calibration of equipment.

§58.61: Equipment used in generation, measurement, or assessment of data and equipment used for facility environmental control shall be of appropriate design and adequate capacity to function according to the protocol and shall be suitably located for operation, inspection, cleaning, and maintenance.

§58.63: Equipment shall be adequately inspected, cleaned, and maintained. Equipment used for the generation, measurement, or assessment of data shall be adequately tested, calibrated and/or standardized.

These and similar regulations and guidelines from other agencies are relevant, but not specific, to computers. With time, an increasing number of computer systems were used by organizations in their laboratory studies, for example, for the control and data evaluation of analytical instruments. With this increase in computerization GMP/GLP monitoring bodies began to show an increased interest in the way such systems were designed, developed, operated, documented and controlled to ensure that compliance with GMP/GLP was maintained.

In October 1982 the US FDA published its first policy guide on computer systems: *Computerized Drug Processing; Input/Output Checking*[1] followed by a second guide published in

December 1982: *Computerized Drug Processing; Identification of 'Persons' on Batch Production and Control Records.*[2]

In 1983 the US FDA's *Blue Book* entitled *'Guide to Inspection of Computerized Systems in Drug Processing'* was published.[3] Although it did not specify in detail how to validate and document computer systems, the book attracted considerable interest throughout the industry.

Four years later the FDA published its *Software Development Activities: Reference Materials, and Training Aids for Investigators.*[4] The document describes the fundamental approaches used in software development including design, implementation, testing, installation, operation and quality assurance, and serves as a guide for the FDA officers who inspect computer systems.

The pharmaceutical manufacturing industry has focused much attention on computer systems' validation, in particular in Good Manufacturing Practices environments, and various groups have studied the problem of computerization and compliance with GLP and cGMP regulations.

In 1983 the US Pharmaceutical Manufacturing Association (PMA) established the Computer System Validation Committee (CSVC) to develop guidelines for the validation of computer systems as used in pharmaceutical manufacturing. After more than two years of research, including two three-day seminars on the validation of computer systems in drug manufacturing in 1984[5] and 1986,[6] results of the committee's work were published in a concepts paper[7] in 1986. The ideas in this paper are considered to be fundamental to the validation of those computer systems used in conjunction with GMP-related activities in the pharmaceutical industry. The terms *computer system* and *computerized systems* were defined and described. The life cycle approach, with functional testing of hardware and software based on specified performance, was proposed to validate computer systems. Recommendations were given on the responsibilities of vendors and users for situations in which the laboratory purchases a computer system from an instrument manufacturer. The paper also includes an extensive glossary on terms used for computer systems validation. Response to the publication of the concepts paper was immediate and positive. Purchasers reported shorter debugging times and systems that functioned correctly at start-up. Many also reported improved communications with vendors and with the FDA.[8]

In 1987 the FDA promulgated the Compliance Policy Guide 7132a.15 *Computerized Drug Processing: Source Code for Process Control Application Software Programs.*[9] The guide appeared to be consistent with the PMA's concept paper: both defined *applications programs* similarly. According to both, purchasers need to possess, and be familiar with application-specific source code, but neither implied that purchasers need possess, or be familiar with, the operating system's source code.

Since 1989 the PMA CSVC has published a series of articles that provide contemporary insight into specific topics such as establishing vendor-user relationships, software categories, system change control and maintenance, information and database management systems, validation documentation, configuration software, quality assurance, training, regulatory compliance, vendor-user relationship, computer systems and software testing.[8,10–13] It was the committee's intent that the papers provide advice but avoid providing directions; consequently, words like *must* and *should* have been avoided.

In 1987 the FDA organized a meeting on computer systems and bioresearch data, inviting individuals from industry, government and academia. The purpose of the meeting was to prepare a reference book of current concepts and procedures in the computer automation of toxicology laboratories that would ensure the quality of computerized data systems. A book, published in 1988, documents their consensus—*Computerized Data Systems for Nonclinical Safety Assessment: Current Concepts and Quality Assurance.*

The issue of computer system validation from the pharmaceutical industry perspective has been addressed by many other authors and organizations. In 1989[14] the United Kingdom Department of Health GLP Monitoring Unit published a document entitled *The Application of GLP Principles to Computer Systems.* The document lays out principles to ensure that as far as possible computers involved in the capture and/or evaluation of data used in support of safety

Table 1.1. Milestones in computer system validation

1982	US FDA publishes first two compliance policy guides on computerized drug processing
1983	US FDA publishes *The Blue Book: Guide to Inspection of Computerized Systems in Drug Processing*
1983	US PMA establishes the Computer System Validation Committee
1985	First widely publicized FDA-483 observations concerning computer systems
1986	PMA concept paper
1987	FDA *Technical report on software development activities*
1987	FDA Compliance Policy Guide: *Computerized Drug Processing; Source Code for Process Control Applications*
1988	Consensus paper: *Computerized Data Systems for Non-clinical Safety Assessments*
1989	UK DOH GLP Monitoring Unit publishes: *The Application of GLP Principles to Computer Systems*
1990	US EPA publishes draft on Good Automated Laboratory Practice
1993	FDA releases draft regulations on the use of electronic records and electronic signatures
1994	The UK Pharmaceutical Industry Computer Systems Validation Forum (PICSVF) releases draft guidelines on *Validation of Automated Systems in Pharmaceutical Manufacture*
1995	The OECD develops a draft paper on *The Application of GLP Principles to Computer Systems*

assessment are developed, tested, introduced and operated in a way that ensures data integrity.

In 1989 Teagarden[15] described the approach that was taken to validate the already existing laboratory system used for data storage and retrieval at the Upjohn company in the United States. The author proposed a stepwise approach to software validation wherein the system was defined, a validation protocol written, the system tested and the documentation reviewed.

In 1990 Stiles[16] wrote an article entitled "GLP and Computerization." The article describes the UK program for GLP and how GLP is applied to computer systems. The author pointed out that system validation with acceptance testing is the most critical issue for a system to comply with GLP. In 1991 K. Chapman published two articles describing the history of validation in the United States.[17,18] Part I covers validation history before computer systems; part II addresses computer system validation. In 1992 Deitz and Herald[19] presented a software development methodology that simplified the PMA's life cycle phases.

In 1994 a subgroup of the UK Pharmaceutical Industry Computer System Validation Forum, chaired by Anthony Margetts of Zeneca, released draft guidelines on *Validation of Automated Systems in Pharmaceutical Manufacture*. The guidelines are intended to increase understanding of current regulatory requirements within the pharmaceutical manufacturing industry and its suppliers. The life cycle model is used to describe the validation responsibilities of vendors and users of automated (computer-related) systems.

Besides the US FDA and the Pharmaceutical Industry, the US EPA paid considerable attention to computer validation. Recommendations are documented in the EPA's *Good Automated Laboratory Practices* chapters 7.6 and 7.9.[20] The document was developed by the EPA's Office of Information Resources Management (OIRM) at Research Triangle Park, N.C. with Rick Johnson as project leader. Chapter 7.6 discusses design, calibration and maintenance issues on automated data collection systems and extends the GLP regulations from general to automated equipment. Chapter 7.9 covers specific software related requirements including the availability of SOPs for proper operation, software validation, functional requirements and historical files for software versions. The *life cycle* approach is recommended to demonstrate the validity of software used in automated systems. Criteria and results for acceptance testing and the names of individuals responsible for the tests should be documented.

Other organizations like the Institute of Electrical and Electronic Engineers (IEEE)[21,22] and the International Organization for Standardization (ISO)[23] published guidelines on software development to ensure a certain software quality for all industries using computer systems in either product development, marketing or services.

Probably the most comprehensive book on computer system and software requirements engineering was published in 1990 by M. Dorfman and R.H. Thayer as an IEEE Computer Society Press Tutorial, *Standards, Guidelines, and Examples on System and Software Requirements Engineering*.[24] The tutorial includes 13 software requirements standards and guidelines for developing, testing and verifying software requirements. It begins with a discussion on the extremely popular *IEEE Guide to Software Requirements Specifications*, developed between 1980 and 1983 by the Software Requirements Working Group of the Software Engineering Standards Subcommittee on Software Engineering of the IEEE computer society. The guide provides a valuable product standard for software requirements specifications. The British Standard[25] relates the specification of the user requirements and the Canadian Standard[26] that of the system design. The *Guideline for Life Cycle Validation, Verification, and Testing of Computer Software*,[27] published by the US National Bureau of Standards (since renamed National Institute of Standards and Technology), provides a methodology for verifying a software requirements specification. It was written as a guide to federal agencies on how to verify, validate and test software products. Included are several methods for verifying, validating, and testing software requirements, design, and code. Software development standards have also been developed by the European Space Agency (ESA)[28] and by the National Aeronautics and Space Administration (NASA).[29]

Independent third parties have developed strategies for computer system validation and offer handbooks for system validation and validation newsletters.[30] Books have been published by different private authors that guide users of computer systems through validation and documentation, for example, *Computer Validation Compliance* by M.E. Double and M. McKendry,[31] *Computer Systems Validation for the Pharmaceutical and Medical Device Industries* by R. Chamberlain,[32]

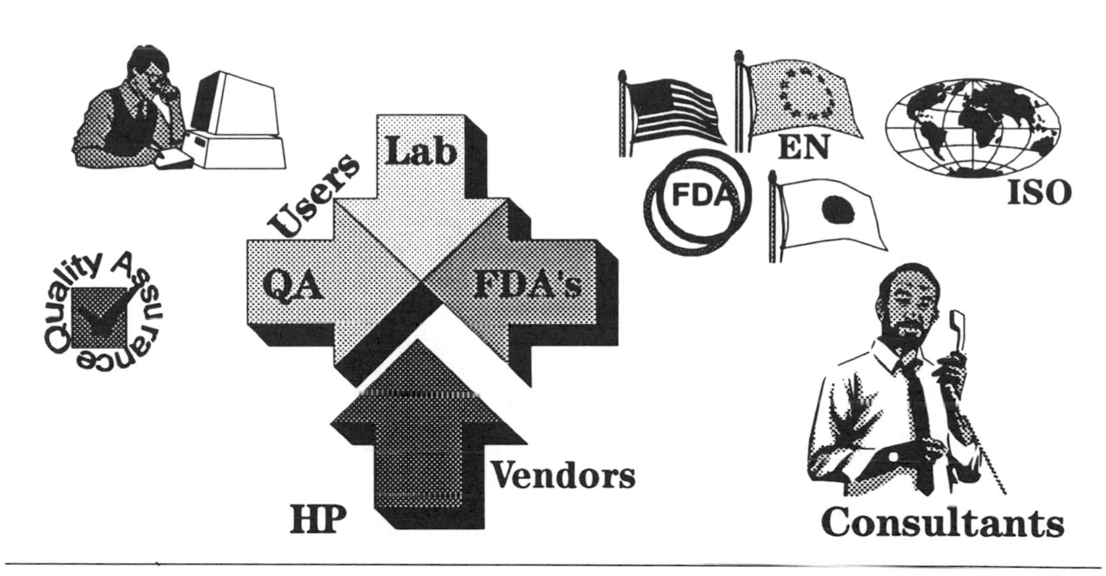

Figure 1.1. Hewlett-Packard works together with users of equipment, regulatory agencies and consultants to get a common understanding on the validation requirements. The goal is to develop products with built-in meaningful automated validation features for users in regulated laboratories.

Good Computer Validation Practices by T. Stokes, R.C. Branning, K.G. Chapman, H. Hambloch and A. Trill,[33] and *Documentation Basics* by C. DeSain.[34]

Despite official guidelines on the development, testing and auditing of software and computer systems, there is still lots of room for interpretation. User firms and field investigators often have different expectations for computer system validation, and even field investigators or different departments within the same firm may not find themselves in agreement. A common understanding amongst all parties involved is of vital importance to avoid frustrations during company-internal and regulatory audits. It is difficult to obtain clear written statements from employees of regulatory agencies when it comes to the details of validation. Sometimes field inspectors publish their expectations and findings on computer system audits. Early in 1993 Tetzlaff[35] published a paper on GMP documentation requirements for automated systems: *FDA inspections of computerized laboratory systems* that included a list of findings from computer system inspections. In a series of articles UK inspector A. Trill reported his findings from inspections of computer systems used in the manufacturing of tablets,[36] while Clark[37] suggested some areas that an FDA investigator might cover during the inspection of an automated manufacturing system.

This book is intended to help readers reach a common understanding on the validation of software and computer-controlled analytical systems.

Precisely which validation efforts are recommended for analytical laboratories depends on the analytical tasks and on quality standards and regulations applied in various countries and companies, and on the type of industry and departments affected by these. Therefore, the book starts with a chapter on quality standards and regulations that have an impact on the validation of analytical equipment and computer systems.

After a chapter on regulation and quality standards, we shall continue with a definition of computer systems and computerized systems. The software life cycle approach, as published by the PMA, EPA, ISO and IEEE and as used by an instrument vendor, is discussed. This book also includes recommendations for the vendor-user relationship as proposed by PMA, US FDA field investigators and OECD. Finally, we give recommendations for retrospective evaluation and validation and examples of test procedures and protocols for acceptance or performance testing of a computer-automated analytical system.

The main focus of the book is on the validation of software and computer systems. Validation of computer-controlled analytical systems also includes testing and calibration of analytical instrument hardware. Strategies for testing and calibration performed as part of the overall system validation process will be discussed in general.

Even though the book uses mostly liquid and gas chromatographs from Hewlett-Packard as examples, the concepts and strategies can be applied to the validation of other computer-controlled analytical systems, such as spectrometers, to other computer systems in analytical laboratories, such as laboratory information management (LIMS) systems and to computerized systems from other vendors.

The author has tried to cover as many validation aspects as possible and make references to the relevant quality standards and regulations in the individual chapters. This does not mean that all recommendations should be followed for all analysis work. The reader should carefully evaluate whether or not the recommendations made in the book are appropriate for his or her work, taking the specific circumstances mentioned above into consideration. Conclusions of the evaluation and their implementation in a laboratory should be part of an overall quality system and documented in a quality manual.

If there is any doubt about whether a specific validation effort should be done, the final answer can be obtained only by asking if the validation effort adds any scientific value.

As discussed earlier regulations and quality standards in general are not specific and leave a lot of room for interpretations for users of equipment as well as for inspectors and auditors. If there is any doubt, the final answer can only be obtained by asking if the validation effort adds any scientific value. For example, it is the author's opinion that there are alternative, and probably better, ways to ensure reliable and consistent analysis data other than to audit well recognized vendors of standard computerized analytical equipment. However, some users are faced with this situation as part of an overall cGMP inspection. Therefore the book tries to give guidelines on how to qualify vendors otherwise and to avoid such vendor audits by collecting the right documentation in advance and making it as efficient as possible just in case it is required.

Last but not least, one should not forget that it is the primary goal of any analyst to generate and deliver analysis data that are scientifically sound, whether they be submitted to a regulatory agency as part of a new drug application (NDA) or delivered to a company-internal or external client. Well-designed, regularly calibrated and maintained equipment and validated methods are prerequisites to achieve this goal and are part of Good Analytical Practice (GAP).

2. Regulations and Quality Standards

Legislation is one of two major forces driving the validation of computer systems. The second and more important reason for validation is to improve software development and maintenance processes: doing it right the first time. As an example, the software development life cycle (SDLC) was not invented by any regulatory agency such as the FDA but by Glenford Myers of IBM in the 1950s.[38] The real benefit of software validation and good validation documentation is in the understanding of the software, which is important for effective and correct software maintenance.

Nevertheless, regulatory requirements play a major role in all validation issues and should therefore be familiar to anybody dealing with computer validations as violation of legislation is a criminal act. The impact of validation on a pharmaceutical company can be immense. If evidence of computer system validation cannot be shown during an FDA audit in a manufacturing facility, the company will receive a warning letter. It has been documented that in the United States[30] a company's stock price drops three to six points when such a warning letter becomes public information, and the impact is even more dramatic should an inspector impose a production and shipment hold pending corrective action. The Pharmaceutical Manufacturers Association (PMA) has estimated that a company can lose between US$50,000 and US$100,000 for every day of delay during the pharmaceutical submission process.[30] In various countries the regulations behind validations in general and behind computer system validation in particular are the (current) Good Manufacturing Practices and the Good Laboratory Practices. The ISO 9000 series of standards provides generic quality standards for development, manufacturing and service. Most frequently used quality standards in chemical testing laboratories are the ISO/IEC Guide 25 and EN 45001. ISO 9000-3 provides guidance on software development, distribution and maintenance. In this chapter the impact of these regulations and standards on the validation of computer systems will be discussed.

(Current) Good Manufacturing Practice (cGMP) Regulations

Good Manufacturing Practice regulations have been developed to ensure that medicinal (pharmaceutical) products are consistently produced and controlled to the quality standards appropriate to their intended use. In the United States Current Good Manufacturing Practices are defined in Title 21 of the US Code of Federal Regulations: 21 CFR 210, Current Good Manufacturing Practice for drugs, general and 21 CFR 211, Current Good Manufacturing Practice for finished pharmaceuticals. Any drug marketed in the US must first receive FDA approval and must be manufactured in accordance with the US cGMP regulations. On account of this, FDA regulations have set an international benchmark for pharmaceutical manufacturing.

In Europe local Good Manufacturing Practice regulations exist in many countries. They are based on the European Union (EU) directive: *EC Guide to Good Manufacturing Practice for Medicinal Products.*[39] This EU GMP directive is necessary to permit free trade in medicinal products between the member countries. EU regulations allow for the marketing of a new drug in the 15 member countries with a single marketing approval. The EU GMP is intended to establish a minimum manufacturing standard for all member states. The EU directive has been widely harmonized with the *Guide to Good Manufacturing Practice for Pharmaceutical Products* as developed under the Pharmaceutical Inspection Convention (PIC).

Specific requirements for computers can be found in section 211.68 of the US cGMP regulations:

- Automatic, mechanical, or electronic equipment, including computers, may be used provided they are routinely calibrated, inspected or checked according to a written program designed to assure proper performance.

- Appropriate controls shall be exercised over computer or related systems to assure that changes in the master production and control records or other records are instituted only by authorized personnel.
- Access to computers shall be controlled to prevent unauthorized changes.
- I/O formulae or other records shall be checked for accuracy.
- Written records of the program and validation data shall be maintained.
- A back-up file of data entered into the computer or related system shall be maintained.

Unlike the US cGMP, the *EC Guide to GMP for Medicinal Products*[39] contains an annex with 19 recommendations that are specific to computers. It includes one paragraph specific to personnel associated with computer system design, operation and maintenance. There are several points that are not covered in such detail by the US cGMP:

- A written detailed description of the system should be produced (including diagrams as appropriate) and kept up to date. It should describe the principles, objectives, security measures and scope of the system and the main features of the way in which the computer is used and how it interacts with other systems and procedures.
- The user shall ensure that software has been produced in accordance with a system of quality assurance.
- When critical data are being entered manually (for example, the weight and batch number of an ingredient during dispensing), an additional check on the accuracy of the record should be made. This check may be done by a second operator or by validated electronic means.
- The system should record the identity of the operators entering or confirming critical data. Any alteration to an

entry of critical data should be authorized and recorded with the reason for the change.

- There should be a defined procedure for the issue, cancellation, and alteration of authorization to amend data, including the changing of personal codes. Authority to alter entered data should be restricted to nominated persons. Any alteration to an entry of critical data should be authorized and recorded with the reason for the change.

- Data should be protected by backing-up at regular intervals. Back-up data should be stored at a separate and secure location.

- Data should be secured against willful or accidental damage by personnel or by physical or electronic means. Stored data should be checked for accessibility, durability and accuracy. If changes are proposed to the computer equipment or its programmes, the above checks should be performed at a frequency appropriate to the storage medium being used.

Detailed requirements for computer systems are specified in the Australian *Code of Good Manufacturing Practice for Therapeutic Goods*.[40] It has a special chapter on the use of computers. The following paragraphs list an extract of the code related to the use of computers.

- The development, implementation and operation of a computer system should be carefully documented at all stages and each step proven to achieve its written objective under challenging test conditions.

- Software development should follow the principles of Australian Standard AS 3563: Software Quality Management System.

- A logical diagram of a schematic for software should be prepared for critical evaluation against system design/requirements criteria.

- A control document should be prepared specifying the objectives of a proposed computer system, the data to be entered and stored, the flow of data, the information to be produced, the limits of any variables and the operating program(s) and test programs, together with examples of each document produced by the program, instructions for testing, operating and maintaining the system and the names of the persons responsible for its development and operation.

- Similarly, where a purchased source code is used or modified, the vendor's attention should be directed to AS 3563. Vendors should be asked to provide written assurance that software development has followed the quality management system of that Standard or of an equivalent system.
- Any change to an existing computer system should be made in accordance with a defined change control procedure.
- Data collected directly from manufacturing or monitoring should be checked by verifying circuits or software to confirm that it has been accurately and reliably transferred.
- The entry of critical data into a computer by an authorized person (e.g., entering a master processing formula) should require independent verification by a second authorized person.
- A hierarchy of permitted access to enter, amend, read, or print out data should be established according to user need. Suitable methods of preventing unauthorized entry should be available.
- The computer should create a complete record ("audit trail") of all entries and amendments to the database.
- The recovery procedure to be followed in the event of a system break down should be defined in writing. The procedure should be designed to return the system to a previous state.

The most extensive documentation on validation of automated systems in pharmaceutical manufacture was published in 1994 by the UK Pharmaceutical Industry Computer

Pharmaceutical Industry Supplier Guidance

Validation of Automated Systems in Pharmaceutical Manufacture

London
March 1, 1994

System Validation Forum as draft guidelines entitled *Validation of Automated Systems in Pharmaceutical Manufacture*.[41] The guidelines take account of the requirements of both the European (EC Guide Annex 11) and US (FDA) regulatory bodies. They make use of existing internationally recognized standards where appropriate and have been developed in consultation with the UK Medicines Control Agency (MCA). The guidelines consist of two parts: a procedure of some 20 pages and appendices of about 100 pages intended to give detailed instruction on implementation. They are not intended to be prescriptive, but rather to be adapted to individual company needs and circumstances and carry no legal or accredited status. They were not designed for retrospective validation of computer systems, but for prospective validation of new systems. They are applicable to all types of new systems including computer-controlled analytical systems and are valid for both bespoke and standard products as well as configurable systems. Rosser[42] gave a broad outline on where the guidelines came from and to where they are supposed to lead; while *Pharmaceutical Technology Europe*[43] has published a condensed form of the introduction to the guidelines.

Good Laboratory Practice (GLP) Regulations

Good Laboratory Practice regulations for assuring the validity of toxicological studies were first proposed by the United States FDA in November 1976 and final regulations codified as part 58 of Chapter 21 of the Code of Federal Regulations in 1978. The United States Environmental Protection Agency (EPA) safety testing of agricultural and industrial chemicals under the Federal Insecticide, Fungicide and Rodenticide Act (FIFRA)[44] and the Toxic Substance Control Act (TSCA)[45] respectively, issued almost identical regulations in 1983 to cover required health and safety aspects. The Organization for Economic Cooperation and Development (OECD) published the principles *Good Laboratory Practice in the Testing of Chemicals*[46] in 1982 which has since been incorporated by OECD member countries into their own legislation. In Europe the European Community (EC) has made efforts to harmonize the laws through council directives on *The Harmonization of Laws, Regulations and Administrative Provisions to the Application of the Principles of Good Laboratory Practice and the Verification of their Application for Tests on Chemical Substances* (1987)[47] and on *The Inspection and Verification of Good Laboratory Practice* (1988, adapted in 1990).[48] To overcome trade differences and enable GLPs to be recognized abroad, bilateral memoranda of understandings (MOUs) were developed.

Currently, most GLP regulations do not have specific requirements that apply to computer and data handling validation. Requirements on equipment design and maintenance are specified in GLP regulations and apply to instruments with and without computers, for example, US GLP regulations, sections 58.61 and 58.63:

- Automatic, mechanical, or electronic equipment used in the generation, measurement, or assessment of data shall be of appropriate design and adequate capacity to function according to the protocol and shall be suitably located for operation, inspection, cleaning and maintenance.
- Equipment used for generation, measurement, or assessment of data shall be adequately tested, calibrated and/or standardized.
- Written standard operating procedures shall set forth in sufficient detail the methods, materials, and schedules to be used in routine inspection, cleaning, maintenance, testing, calibration and/or standardization of equipment and shall specify remedial action to be taken in the event of failure or malfunction of equipment.
- Written records shall be maintained of all inspection operations.

In 1989 the United Kingdom Department of Health GLP Monitoring Unit published a document entitled *The Application of GLP Principles to Computer Systems*[14] outlining how inspectors will approach the examination of computer systems in laboratories conducting human health and environmental safety studies. It is also of assistance to laboratory management in identifying the aspects of GLP that should be considered when installing and commissioning computer systems. The document includes a chapter on the "Interpretation of the GLP Principles for Computer Systems" with paragraphs on:

- Identification and definition of the system
- Control procedures for programs within a system, for applications software and for security
- Archives
- Quality assurance
- Staff training

In Japan the Ministry of Health and Welfare (MOHW)[49] has an annex to its GLP regulations that is specific to computer systems. The guide includes specific recommendations for the prospective validation of computer systems that have been developed in-house as well as for systems purchased from a vendor and in addition provides guidelines for the retrospective validation of existing systems. Also included is a section on systems where the laboratory has programmed additional software to further customize software purchased from a vendor: "In cases where the testing facility developed a part of the system for the purpose of supporting the vendor-supplied software, check that this part of the system has been documented and approved."

The OECD GLP consensus document number 5 on *Compliance of Laboratory Suppliers with GLP Principles*[50] includes a chapter on computer systems in which the responsibilities of the software user (including software obtained from an external supplier) are defined:

- The user should ensure that all software obtained externally has been provided by a recognized supplier. Implementing ISO 9000 is considered to be useful for a software supplier.
- It is the responsibility of the user that the software program has been validated. The validation may be undertaken by the user or the supplier, but full documentation of the process must be available.

- It is the responsibility of the user to undertake an acceptance test before using the software program. The acceptance test should be fully documented.

Good Clinical Practice (GCP) Regulations

Computers are widely used in clinical laboratories. In larger laboratories, the handling of the sample from receipt to the final report is usually totally automated. Some countries have included chapters on the use of computers in their GCP regulations or guidelines. For example, in Europe the Commission of the European Communities has published a guide entitled *Good Clinical Practice for Trials on Medicinal Products in the European Community.*[51] Chapter 3 includes requirements for the use of validated programs and computerized systems, audit trail of data and the verification of transformations:

- Computerized systems should be validated and a detailed description of their use be produced and kept up to date.
- For electronic data processing, only authorized persons should be able to enter or modify data in the computer and there should be a record of changes and deletions.
- The sponsor must use validated, error free data processing programs with adequate user documentation.
- The sponsor must ensure the greatest possible accuracy when transforming data. It should always be possible to compare the data printout with original observations and findings.
- Archived data may be held on microfiche or electronic record, provided that a back-up exists and that hard copy can be obtained from it if required.

In the United States there is no single equivalent GCP document. A series of Good Clinical Practices guidelines and regulations were published in 1977, 1981, 1987 and 1991.[52–56] While none of the US GCP documents have specific chapters on the use of computerized systems, this does not mean that the FDA does not care about the validation of computer systems used in clinical trials. The common practice is to refer to documents published by the FDA on validation and inspection of computer systems, for example, inspection and compliance policy guides.[3,9]

Good Automated Laboratory Practice (GALP)

EPA
Good Automated Laboratory Practice
Draft Recommendations for Ensuring Data Integrity in Automated Laboratory Operations with Implementation Guidance

The most extensive document on the use of automated (computer) systems published by governmental agency are the US. EPA's draft recommendations for Good Automated Laboratory Practice.[20] They were developed in response to the need for standardized laboratory management practices.

The *Good Automated Laboratory Practices*, first published as a draft in December 1990 by the US Environmental Protection Agency (EPA), was widely distributed, and comments were gathered. The document gives recommendations on how to operate laboratory information management systems (meaning automated data processing, telecommunications, storage, and retrieval technology) in an EPA regulated environment to ensure integrity, traceability and safety of the data. Besides official recommendations, the document also includes extensive interpretation of the guidelines.

The document has two parts: 19 pages containing 83 concisely written recommendations and more than 200 pages of implementation notes and guidance, discussing each of the 83 recommendations in detail. Implementation notes include explanations of the recommendations, making the document an invaluable asset. While in the Good Laboratory Practice regulations usually there is little said on implementation, most of the GALP document is implementation guidance using practical application examples. There are also 'special consideration' paragraphs that again are very helpful.

The recommendations found in the document are not necessarily limited to studies controlled by the EPA and may be used as a yardstick by any laboratory to measure the accuracy and integrity of data generated through automated laboratory operations.

GALPs can be applied whenever an automated data collection system is used, regardless of the size and scope of the storage computer.[57] Although the GALP manual specifically mentions Laboratory Information Management Systems (LIMS) applications it is the understanding of the EPA that the same principles apply equally well to data acquisition from computerized analytical instrumentation, from an automated HPLC or GC/MS system, for example.[57]

Currently published as a draft, the recommendations are built on six principles:

1. Data: assure integrity of all entered data.
2. Formulae: assure that formulae and algorithms are accurate and appropriate.
3. Audit trail: track data/edits to responsible person.
4. Change: provide an adequate change control procedure.
5. SOPs: use appropriately documented procedures.
6. Disaster: provide alternative plans for failures and unauthorized access.

The GALP recommend the appointment of a person primarily responsible for the computer system: The laboratory shall designate a computer scientist or other professional of appropriate education, training or experience or combination thereof as the individual primarily responsible for the automated data collection system(s).

The person is responsible for:

- Sufficient qualified personnel for design and use of the system
- Security
- Standard operating procedures and other documentation
- Change control
- Data recording
- Problem reporting
- GLP/GALP compliance

In general, requirements for computer operation, data traceability and security are much more specific than stated in GLP regulations.

Standard operating procedures should be available for:

- Maintaining security (physical security, access to the system and its functions and restricting installation of external programs/software).
- Entry of data and proper identification of the individual entering the data.
- Verification of manually or electronically input data.
- Interpretation of error codes or flags and the corrective action to follow when these occur.
- Changing data and methods for execution of data changes. Any time data is changed, for whatever reason, the date of the change, reason for the change and individual making the change must be indicated, along with the old and the new values of the data elements that have been changed.
- Data analysis, processing, storage, and retrieval.
- Back-up and recovery of data.
- Maintaining hardware.
- Electronic reporting, if applicable.
- Availability of documents such as standard operating procedures and operating manuals.

Other software, instrument or data specific recommendations include:

- For software purchased from a vendor an attemption should be made to obtain the documentation from the vendor, otherwise it should be reconstructed in-house, to the degree possible.
- Detailed descriptions on the software's functionality must be developed and available in the laboratory. For software

purchased from vendors it is expected that the vendor provides these descriptions.

- Written documentation on software development standards must exist including programming conventions.
- All formulae or algorithms used for data analysis, processing, conversion or other manipulations should be documented and retained for reference and inspection. For purchased software, formulae and algorithms may be obtained from vendor-provided documentation. For most software currently in use, it is probable that formulae and algorithms will have to be abstracted.
- Acceptance criteria should be specified for new or changed software. Written procedures should be in place for acceptance testing including what should be tested, when and by whom. Test results should be documented, reviewed and approved.
- When software is changed, written documentation must exist with instructions for requesting, testing, approving and issuing software changes.
- Procedures must be in place that document the version of software used to create or update data sets. An audit trail must be established and retained that permits identification of the software version in use at the time each data set was created.
- A historical file of operating instructions, changes and version numbers shall be maintained. All software revisions, including the dates of such revisions, shall be maintained within the historical file. The laboratory shall have appropriate historical documentation to determine the software version used for the collection, analysis, processing or maintaining of all data sets on automated data collection systems. (This will allow the laboratory to audit trail data sets with the software version used to generate them.)
- The individual responsible for direct data input shall be identified at the time of input. (This can be accomplished by forcing the operator to enter a password into the computer system before any data entry can be made.)

- If data are entered directly from analytical instruments, the instrument transmitting the data shall be identified, together with the time and date of transmittal. (This can be accomplished by entering the instrument's serial number along with a date and time stamp into each data set transmitted to the computer.)

- Any change in automated data entries shall not obscure the original entry, shall indicate the reason for the change, shall be dated and shall identify the individual making the change.

- The accuracy of data must be verified, no matter if data have been entered manually or if they have been electronically transferred from an analytical instrument. Data validation methods include double-keying of manually entered data and blind rekeying of data entered automatically.

- All raw data, documentation and records generated in the design and operation of automated data collection systems shall be retained.

- Written procedures, such as company policies or SOPs on the definition of computer resident raw data, must be retained for inspection or audit.

- Written descriptions of the hardware and software used for the data collection, processing and archiving should exist. Hardware descriptions and general software descriptions are usually provided by the vendor, but system configurations can be documented in-house.

- Acceptance test records obtained from new and modified software tests must be permanently retained. Records must include the tested item, method of testing, the date(s) the test was performed, and the individuals who conducted and reviewed the tests.

- Records of periodic laboratory inspections conducted by the QA unit shall be retained. Equipment inspections include hardware, software and computer resident data.

Quality Standards and Guidelines

Most chemical analytical laboratories already have, or are in the process of implementing, a quality management system to improve the quality, consistency and reliability of data. A documented quality system is also a prerequisite for obtaining accreditation or registering for a quality standard such as ISO 9001, 9002 or 9003.

EN 45000 Series and ISO/IEC Guide 25

The European Norm EN 45001:1989 General criteria for the operation of testing laboratories[58] and the ISO/IEC Guide 25 General requirements for the competence of calibration and testing laboratories[59] are frequently used as guides in establishing a Quality System in chemical testing laboratories, and both documents may be used as a basis for laboratory accreditation.

The EN 45001 and the ISO/IEC Guide 25 both include a chapter on equipment similar to the equipment sections found in GLP and cGMP regulations.

Section 5.4.1 of EN 45001 includes several paragraphs on electronic data processing and test methods and procedures that discuss the following:

- Where results are derived by electronic data processing techniques, the reliability of the system shall be such that the accuracy of the results is not affected.
- The system shall be able to detect malfunctions during execution and take appropriate actions.

Section 10.7 of the ISO/IEC Guide 25 includes several paragraphs on the use of computers. Where computers or automated equipment are used for the capture, processing, manipulation, recording, reporting, storage or retrieval of calibration or test data the laboratory shall ensure that:

- Computer software is documented and adequate for use.
- Procedures are established for protecting the integrity of data; such procedures shall include, but not be limited to,

integrity of data entry or capture, data storage, data transmission and data processing.

- Computer and automated equipment is maintained to ensure proper functioning and provided with the environmental and operating conditions necessary to maintain the integrity of calibration and test data.

- It establishes and implements appropriate procedures for the maintenance of security of data including the prevention of unauthorized access to, and the unauthorized amendment of, computer records.

More specific information can be found in the *EURACHEM Guidance Document No. 1/WELAC Guidance Document No. WGD 2: Guidance on the Interpretation of the EN 45000 Series of Standards and ISO/IEC Guide 25.*[60] Section 10 on the *Use of Computers* includes several paragraphs:

- For the purpose of validation of computers used in chemical testing it is usually sufficient to assume correct operation if the computer produces expected results when input with known parameters.

- It is possible to validate a computer based data processing system by running it in parallel to an analog chart recorder or non-computer based integrator, calibrated against a traceable signal generator.

- (Computer) system validation should be achieved by validation of individual components plus an overall check on the dialogue between individual components and the controlling computer.

- Such (computer) systems will normally be validated by checking for satisfactory operation (including performance under extreme circumstances) and establishing the reliability of the system before it is allowed to run unattended.

- Where possible the controlling software should be tailored to identify and highlight any malfunctions and tag associated data.

- Electronic transfer of data should be checked to ensure that no corruption has occurred during transmission. This can be achieved on the computer by the use of verification files, but wherever practical the transmission should be backed-up by a hard copy of data.

Appendix C of the same document gives further guidance on the use of computers:

- The laboratory should document any special procedures relating to security, power failure, file management (including archiving, file repair, file back-ups), validation and training.

- The laboratory should take appropriate measures to safeguard the integrity of computers, software and associated data.

- The laboratory should prevent unauthorized access to computers. Suitable measures will include secure areas, keyboard locks and password routines, voiceprints and fingerprints.

- All software and subsequent updates should be authorized before use, having first been screened for viruses and validated.

- Ideally, software should only be obtained from reputable sources or written in-house.

- Where software is written in such a way that the content of a file is intended to be easily accessible for changing, there is normally a mechanism as part of the file management system that indicates the date of the last change to a particular file and in some cases indicates the complete history for each file.

- Software is often imperfect. It must be fully documented. Wherever possible a record of known faults ("bugs") should be obtained from the supplier and any effect that these may have on the day-to-day running of the software should be established.

- For each computer the proposed use should be defined so that the degree of validation necessary may be established.
- Where software has been updated, a record should be kept of the revision history.
- Where sample results are archived, it is necessary to store all of the information required to reproduce the original answers. In addition to the raw data files this will include any associated data-processing files and, if appropriate, the relevant version of the operating software. In extreme cases it may be necessary to retain obsolete hardware in order to run the archived software.
- Computers, whatever their type, suffer from the 'black-box' syndrome: an input is made at one end, an answer is produced at the other. Because what happens inside cannot be seen, it must be assumed that the box is functioning correctly. For the purpose of validation it is usually acceptable to assume correct operation if the computer produces expected answers when input with well-characterized parameters.
- The degree of validation necessary depends on the exact use of the computer.

NAMAS Accreditation Standard

NAMAS, the UK National Measurement Accreditation Service has a paragraph on computer systems in its accreditation standard M10[61]: "Where computers or automated test equipment are used for the collection, processing, recording, reporting, storage or retrieval of calibration and test data the laboratory shall ensure that, where applicable, the requirements of 6 of this Standard are met. The laboratory shall, wherever possible, ensure that computer software is fully documented and validated before use." In addition NAMAS has developed *A Guide to Managing the Configuration of Computer Systems (Hardware, Software and Firmware) Used in NAMAS Accredited Laboratories.*[62] The document outlines methods for the management and control of the configuration of computer systems to ensure that the requirements expressed more generally in M10 are met.

ISO 9000 Series of Quality Standards and ISO 9000-3

The Quality Standards ISO 9001 to ISO 9003 cover the requirements for a generic quality system for a two-party contractional situation with an assessment made by a third party. The standards are not specific to laboratory work nor to computer systems or software. Recognition that the development, manufacturing and maintenance processes of software are different to most other products led to the issuance in 1991 of ISO 9000-3: *Guidelines for the Application of ISO 9001 to the Development, Supply, and Maintenance of Software.*[23] It provides additional guidance for quality systems involving software products and deals primarily with situations where specific software is developed, supplied and maintained according to a purchaser's specification as part of a contract.

TickIT

Software development and maintenance activities can be formally assessed using the TickIT scheme through the application of ISO 9000-3. The scheme was investigated by the Department of Trade and Industry in the United Kingdom. The TickIT guide[63] comprises 172 pages with chapters on:

- Introduction
- ISO 9000-3: 'Guidelines for the Application of ISO 9001 to the Development, Supply and Maintenance of Software'
- Purchaser's guide
- Supplier's guide
- Auditor's guide (includes the European IT Quality System Auditor Guide)

The guide complements ISO 9000-3 by providing additional guidance on implementing and auditing a Quality Management System (QMS) for software development and support.

3. Definitions: Computer Systems, Computerized Systems and Software

The terms *computer system* and *computerized systems* have been very well defined.[7] The computer system as used in an analytical laboratory consists of computer hardware, peripherals and software to perform a task. Software includes operating environments such as Microsoft® MS-DOS® and Microsoft Windows™ and the standard applications software, for example, the Hewlett-Packard ChemStations.

The computer is closely associated with instrument hardware, such as a balance or a gas chromatograph. The software on the computer not only controls the instrument, a temperature gradient for instance, but also acquires data from the instrument, for example, signal or spectral data from a mass spectrometer. The entire analytical system consisting of computer hardware, software and analytical hardware is defined as a *computerized system.* A computer-related system includes, in addition, the physical and operational environment of the computerized system, for example, interfaces to a second computerized system, to a database or to other equipment.

Figure 3.1 shows an example of a computerized, automated HPLC system comprising an HPLC instrument with autosampler, pump, column and detector plus computer hardware and software, plus documentation on how to operate and test the entire system. J. Guerra[64] described a computer system consisting of a workstation (computer hardware and software) and laboratory equipment connected to the computer either through an integrator or directly through an RS 232 interface.

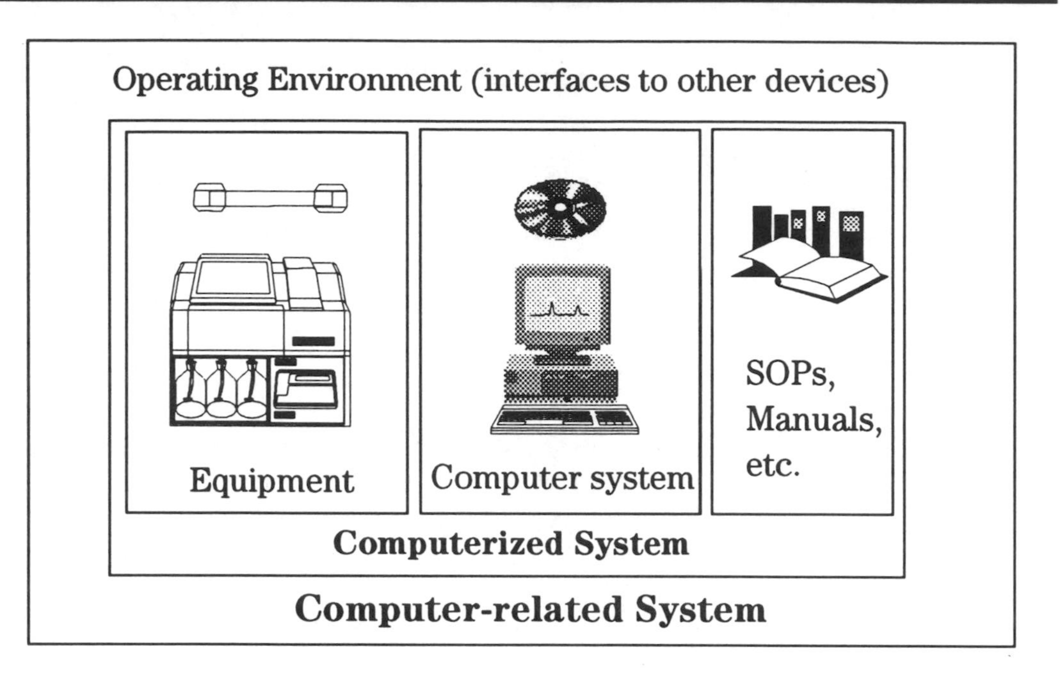

Figure 3.1. Computer-related system

Other examples of automated systems in a laboratory include laboratory data systems with data acquisition from a large number of instruments through A/D converters, laboratory information management systems (LIMS) for information handling and data archiving and a server for multitechnique data review and reporting. A LIMS has been described by Garfield[65] as "A database tailored to the analytical laboratory so that it can handle data generated by the analysis of samples and integrate sample information with results obtained from analytical instruments, thereby reducing administrative tasks and increasing the production of the final reports."

On a computer used for analytical instrument control we can generally find three different software categories:

1. *System software,* such as operating software DOS® or UNIX®, supplied by software firms, e.g., Microsoft®. This also includes such utilities as device drivers and file management.

2. Canned *standard application software,* for example, off-the-shelf chromatography software, generally supplied by an instrument vendor.

3. *User specific application software,* written by the user or by a third party for a specific user to meet specific needs in the user's laboratory.

Microsoft® MS-DOS®, Windows™, OS/2® and UNIX® are currently the most frequently used operating systems and user interface environments. They are supplied with the computer in a machine-executable format that cannot be modified by the user and is not unique to any one user's system. Examples of standard applications software are the ChemStations (DOS Series) from Hewlett-Packard, which are available for instrument control and data evaluation of GC, LC, MS, UV-visible spectrophotometers, CE and sample preparation systems. This software is sometimes referred as canned user configurable[18] because the user enters his or her own application specific information. The software is also supplied with the computer in a machine-executable format that cannot be modified by the user and is also not unique to any one user's system.

- **System software**
  independent of specific application
- **Canned standard applications software**
  user defines conditions, operating and reporting parameters
- **User specific applications software**
  for specific user needs

Examples

Operating systems
MS-DOS, UNIX

HP ChemStations

MS Excel MACRO
or
HP ChemStation
MACRO

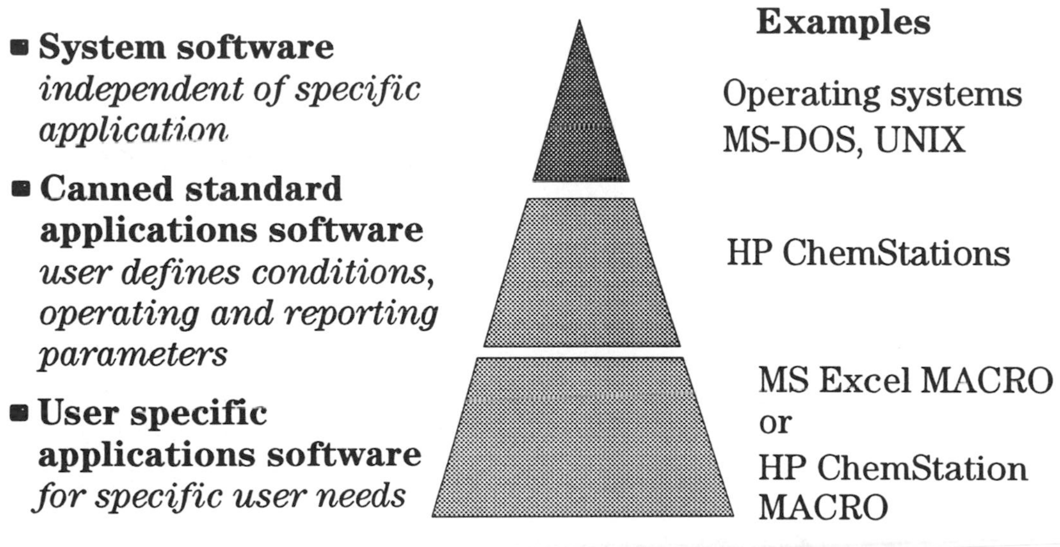

Figure 3.2. Types of computer software in analytical laboratories and examples

Examples for user specific application software are ChemStation or spreadsheet macro programs for customized data evaluation and reporting.

Even though the user has the ultimate responsibility to ensure that the software is validated,[35,50] in practice validation of software from different categories is done at different sites. Operating systems are expected to be validated by the copyright owner of the software, Microsoft®, for example. Normally, chromatography software vendors validate the complete system including structural testing (so-called white box testing), and functional testing (so-called black box testing), which comprises acceptance and performance testing. The user of such software performs only the functional, acceptance and performance testing within the laboratory environment. If the chromatography software needs to be customized for a special report or data evaluation package, the user may write a program using a set of commands included in the standard software. In this case, the user has to validate this part of the overall software including structural, functional and performance testing.

Validation of a computer-controlled analytical system requires calibration and/or testing of the individual instrument modules, the interface to the equipment and the computer, the computer connected to the instrument as well as the overall system with instruments connected to the computer. Instrument hardware should be calibrated and tested for performance, for example, the wavelength accuracy of a UV-visible HPLC detector or the precision of peak areas of a gas chromatograph.

4. Validation Processes in the Analytical Laboratory

Accurate and reliable analytical data in laboratories can only be obtained with validated equipment, methods and processes for data validation. Regulatory agencies and independent third party auditors expect equipment and laboratory processes used during the generation of analytical data to be validated. This is the case for all types of equipment, including computerized analytical instrumentation.

Definitions of Validation

The term *validation* has been defined in literature by many different authors. Although the wording is different, the sense is always the same: (a) specify and implement, (b) test if the specifications are met and (c) document.

In the area of software engineering, the US Institute of Electrical and Electronic Engineers (IEEE) gave a definition of software validation as: "The process of evaluating software at the end of the software development to ensure compliance with software requirements."[21,22]

This definition of validation is not commonly used by the pharmaceutical industry because it is today's understanding that the validation process starts at the beginning of the life cycle of the software and ends when it is no longer in use.

One of today's commonly accepted definitions of validation can be found in the guideline *General Principles of Validation* from 1987.[66] "Establishing documented evidence which provides a high degree of assurance that a specific process will

Definition

Establishing documented evidence which provides a high degree of assurance that a specific process will consistently produce a product meeting its predetermined specification.

Source: FDA guidelines on General Principles of Validation, March 1986

Continuous process

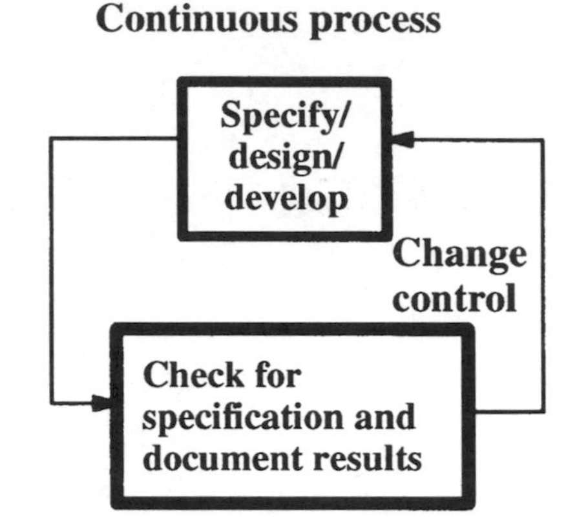

Validation means nothing else than well-organized, well-documented common sense (Ken Chapman, 1985).

Figure 4.1. Principle and definitions of validation

consistently produce a product meeting its predetermined specifications and quality attributes."

This definition is very well thought out and each word has a special significance. Most important in this definition are the words documented, high degree of assurance, specific process, consistently, and predetermined specifications.

- **documented:** Validation requires a thorough documentation. Everything that is not documented is considered not to be done.
- **high degree of assurance:** The assumption is that a large software package as used in complex computerized systems is rarely free of errors.
- **specific process:** The overall validation of software is process, not product, related. For example, the development and testing activities before releasing the software for manufacturing are validated once for a series of products characterized by the serial number. Some subparts of validation, such as the qualifications (installation, operational, performance), are product specific and have to be done for each system.

- **consistently:** Validation is not a one-time event. The performance of the computer system has to be controlled during the entire life of the product.

- **predetermined specifications:** Validation activities start with the definition of specifications. The performance of the computer system is then verified against these specifications. Acceptance criteria must be defined prior to testing.

Figure 4.2 illustrates the validation time line of a water treatment system. The validation model is based on a concept paper developed by the US PMA Deionized Water Committee.[67] It demonstrates that validation is not a one-time event but an ongoing process starting with the definition and design of the product. The paper also uses terms such as Installation Qualification (IQ), Operational Qualification (OQ) and Performance Qualification (PQ). Meanwhile these terms are also used for the validation of analytical equipment in pharmaceutical manufacturing. They will be explained later.

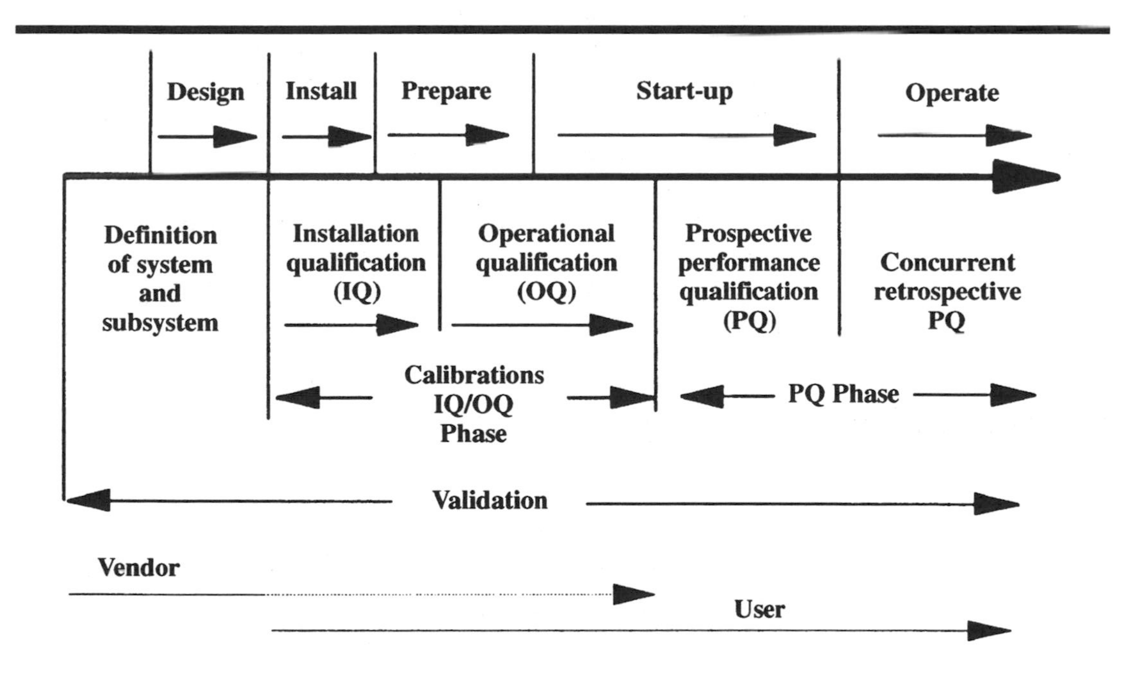

Figure 4.2. The validation time line, originally developed for the validation of a water treatment system, now frequently applied for the validation of analytical equipment used in pharmaceutical manufacturing.[15]

The *EURACHEM/WELAC Guidance on Interpretation of the EN 45000 Series of Standards and ISO/IEC Guide 25*[60] defined validation of data and equipment in appendix C1.11 as "The checking of data for correctness, or compliance with applicable (of data processing) standards, rules and conventions. In the context of equipment rather than data, validation involves checking for correct performance etc."

The US Pharmacopeia[68] defined validation of analytical methods as "The process by which it is established, by laboratory studies, that the performance characteristics of the method meet the requirements for the intended applications."

Many laboratory managers associate validation with increased workload in the laboratory or increased paperwork, but validation is essentially nothing new. Ever since the development of analytical instrumentation and methods, statistics have been used to prove the proper functioning, reliability and precision of the equipment and methods. Firms followed software development standards and used the software development life cycle long before regulatory agencies requested the validation of computer systems. New to most existing validation procedures is the disciplined planning of validation and documentation of all validation steps, including testing. This also concurs with a definition of validation by Ken Chapman[17]: "In today's pharmaceutical industry, whether you are thinking about a computer system, a water treatment system, or a manufacturing process, validation means nothing else than well-organized, well-documented common sense."

Validation vs. Verification, Testing, Calibration and Qualification

There is still considerable misunderstanding on the differences between testing, verification, and validation. The illustration in Figure 4.3 should help to clarify this.

Testing

Testing has been defined in the ISO/IEC Guide 25 as: "A technical operation that consists of the determination of one or

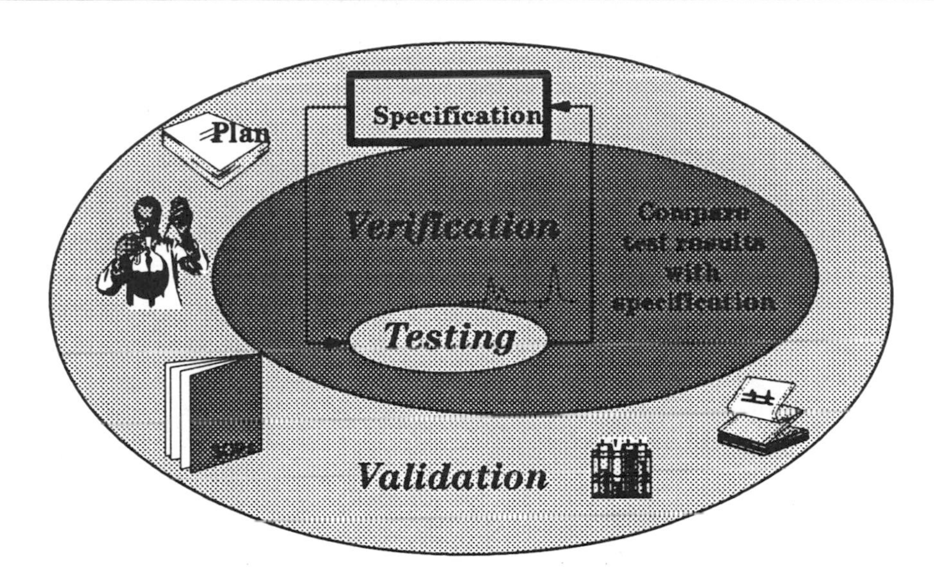

Figure 4.3. Testing, verification and validation

more characteristics or performance of a given product, material, equipment, organism, physical phenomena, process according to a specified procedure."

Instrument testing is the process of executing experiments to measure its performance characteristics following documented procedures. Examples are the measurement of the baseline noise of a detector, the precision of the injection volume of an injector or the precision of a flow rate. Requirements for testing are test conditions and standard operating procedures with clear instructions on how to do the tests and how to evaluate the results.

Verification

The ISO/IEC Guide 25 defines verification as the "Confirmation by examination and provision of evidence that specified requirements have been met." Performance Verification (PV) of analytical instrumentation is the process of comparing the test results with the specification. It includes testing and requires the availability of clear specifications and acceptance criteria. Examples are the same as for testing. The

Table 4.1. Documentation that should be available or generated before, during and after testing, verification and validation. Testing is included in verification and calibration. Verification and calibration are part of the qualification. Qualification is part of the validation.

	Documents for start and execution	Documents generated during execution
Testing	☑ test conditions ☑ SOPs for execution ☑ forms for test results	☑ sheets with test results ☑ logbook entries
Verification	☑ SOPs for execution ☑ specifications ☑ acceptance criteria	☑ sticker to affix to the instrument ☑ declaration of performance verification ☑ logbook entries
Calibration	☑ SOPs for execution results ☑ list with expected calibration results	☑ sheets with calibration ☑ sticker to put on the instrument ☑ declaration of calibration ☑ logbook entries
Qualification (installation, operational, performance)	☑ templates for installation, operational, and performance qualifications	☑ qualification protocols with results ☑ logbook entries
Validation	☑ validation plan ☑ procedures for execution	☑ validation report

verification process ends with the generation and sign-off of a 'Declaration of Conformity' of the instrument to specifications. Besides that, a sticker should be affixed to the instrument with the date of the last successful performance verification and the next scheduled performance verification.

Calibration

The ISO/IEC Guide 25 defines calibration as "The set of operations which establish, under specified conditions, the relationship between values indicated by a measuring instrument or measuring system and the corresponding known values of the measurand."

An example for a calibration procedure in an analytical instrument is the measurement and adjustment of the wavelength accuracy in an HPLC UV-visible detector's optical unit. Calibration is frequently confused with testing and performance verification. The differences become quite clear when looking at the precision of the peak area: This can be tested and verified against a previously defined specification, but it cannot be calibrated. Sometimes, accurate calibration has a direct impact on performance, for example, a UV detector with incorrect wavelength calibration may cause detection limits and the detector's linearity to deteriorate. Calibration is sometimes used interchangeably with standardization. Calibration normally means to check against known standards, whereas standardization usually means to make uniform.[69] For some equipment the term *calibrated* is more appropriate, and for other equipment the term *standardized* is better.[69]

Qualification

The term *qualification* has been defined by the US PMA's Computer System Validation Committee for installation, operation and running of the system under workload for a specific application. Qualification is part of validation and is unit specific. While the overall validation phases associated with development are process specific and have to be done for each version of the software only once, the qualifications have to be done for each unit. Three qualifications have been defined: installation qualification (IQ), operational qualification (OQ) and performance qualification (PQ).

Validation

While computer system validation includes calibration, testing, verification and qualification activities it actually begins with the definition of the requirements and ends with the removal of the instrument from the laboratory. Different to qualification, calibration, and testing, validation is frequently process or project specific and not related to a specific unit as characterized by a serial number. All activities are listed in a validation plan that should be developed and followed for each project. Validation activities are usually shared between the supplier (process related design and development validation) and the user (product related qualification) of an instrument. For new systems, validation is done prospectively based on a validation plan, starting with design and development. For existing systems the validation is done retrospectively, based on the review and evaluation of historic information.

Validation Steps in the Laboratory

Validation efforts in an analytical laboratory should be broken down into separate components addressing the equipment, the analytical method run on that equipment, the analytical system and finally the analytical data. The various validation activities in an analytical laboratory are illustrated in Figure 4.4.

Analytical equipment hardware should be validated prior to routine use and, if necessary, after repair and at regular intervals. Computer systems should be validated during and at the end of the development process and, if necessary, after software updates. Computer system validation includes the validation process during software development at the development site and the qualifications of the individual products at the user's site. Computer hardware and firmware are validated during their development. Specific computer hardware products are qualified in the user's laboratory as part of the system qualifications and whenever the application is successfully executed and tested.

Method validation covers testing of significant method characteristics, for example, limit of detection, limit of quantitation, selectivity, linearity and ruggedness. If the scope of the method is that it should run on different instruments, the method should also be validated on different instruments. Only when it is clearly specified that the method will always run on the same instrument, can validation efforts be limited to that instrument. Methods should be validated at the end of method development prior to routine use and whenever any method parameter has been changed.

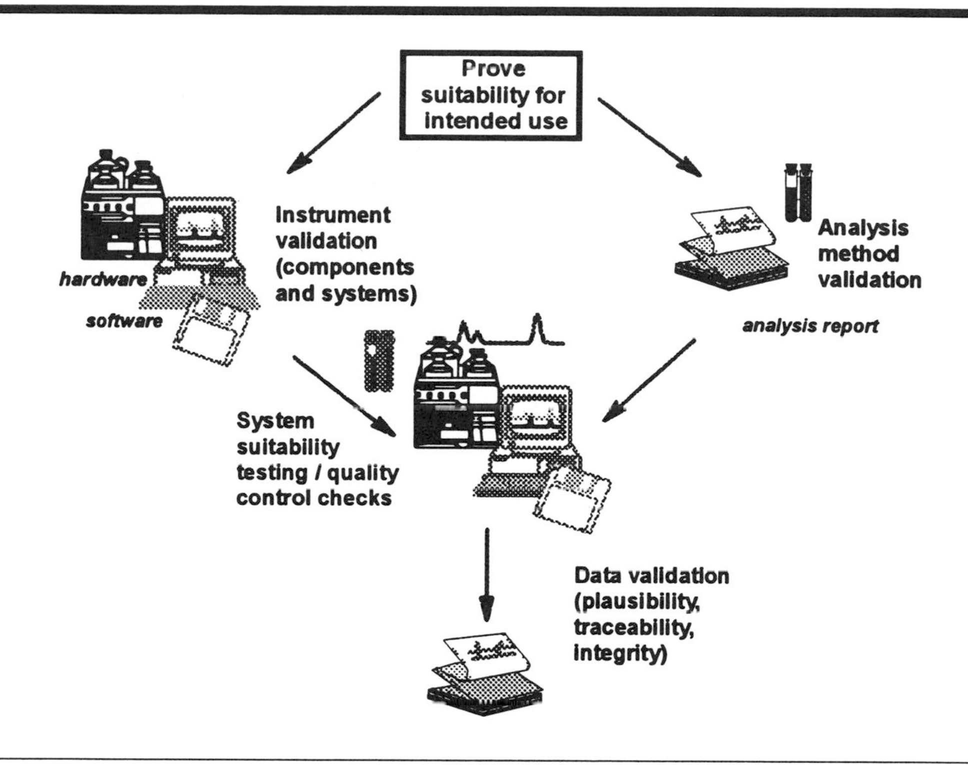

Figure 4.4. Validation in the analytical laboratory

A system combines instrument, computer and method. In chromatography it also includes a column and reference material for calibration. This validation, usually referred to as system suitability testing, tests a system against documented performance specifications, for the specific analytical method. Analytical systems should be tested for system suitability prior to and during routine use, practically on a day-to-day basis.

When analyzing samples the data should be validated. The validation process includes documentation and checks for data plausibility, data integrity and traceability. A complete audit trail that allows the final result to be traced back to the raw data should be in place. During routine analysis, quality control samples with known amounts should be analyzed in between the unknown samples.

Other tasks are equally important in ensuring reliable and accurate data: all laboratory work staff should be adequately qualified, and their qualifications should be documented. Standards for instrument calibration and quality control checks should be checked following documented plans.

In this book validation of the computer-controlled equipment including analytical instrument hardware, the computer system used to control the hardware, the data acquisition process, method validation and data evaluation will be discussed.

5. Life Cycle Approach for Validation of Software and Computer Systems

Software development often takes several years and as Chapman and Harris[8] pointed out, it is impossible to ensure a certain quality standard simply by testing the program at the end of its development process. Quality cannot be designed into the software after the code is written; it must be designed and programmed into software prior to and during its development by following written development standards, including the use of appropriate test plans and test methods.

Software validation differs from hardware validation in that it is harder to specify absolute performance criteria for software and more difficult to define software tests. On the other hand, software has the advantage that it will not physically degrade over time as a detector lamp would, for example. Software failures are design or implementation failures: they are present from the day the software is installed and will remain until found and corrected. Usually they only become evident under certain combinations or circumstances. This makes it impossible to detect all errors by testing. The validation of software should be occurring at the same time as the development.

The product life cycle or system development life cycle (SDLC) approach has been widely accepted to validate software products. This approach to system development and validation is common in many engineering and manufacturing fields and has been standardized by the American National Standards (ANSI), US Department of Defense (DOD), NASA,[29] IEEE,[22] the Pharmaceutical Manufacturers'

User requirement specifications

Product design and development

Testing

- **Structural testing (white box)**
- **Functional testing (black box)**

Installation/Operation

- **Operator Qualification**
- **Installation Qualification**
- **Operational Qualification**
- **Performance Qualification (Ongoing control)**

Figure 5.1. Steps involved in software and computer system validation

Association (PMA),[7] the UK Pharmaceutical Industry Computer Systems Validation Forum (PICSVF),[41] the European Space Agency (ESA),[28] the US Environmental Protection Agency (EPA),[20] and the International Organisation for Standardization (ISO).[23] The development process is divided into phases:

☑ requirements and specifications are set

☑ design and implementation follow, with code generation and inspection

☑ subsystems are tested, then integrated into a system and tested as a system

☑ the system is installed and qualified before it can be put into routine use

☑ history of changes is maintained and recorded

Each phase is completed, reviewed and approved by management before the subsequent phase is started.

The product life cycle approach can be applied to all types of software and computer system projects. For a new system, the

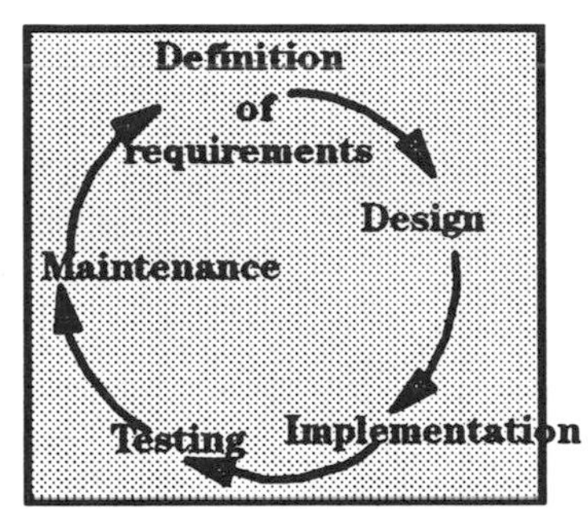

- Quality cannot be built into a product by testing. It must be designed into the product
- Divide the development into phases.
- Complete, review and approve each phase before starting the next phase.

The concept has been standardized by ANSI, NASA, IEEE, ESA, BSI, PMA, ISO.

ANSI = American National Standards Institute, IEEE = Institute of Electrical and Electronic Engineers,
NASA = National Aeronautics and Space Administration, ESA = European Space Agency,
BSI = British Standards Institute, PMA = Pharmaceutical Manufacturing Association,
ISO = International Organization for Standardization

Figure 5.2. Concept of the software development life cycle

life cycle and the validation begins at the point of initial definition and design and ends with the termination of the system. For computer systems already in use, the validation begins with a current definition of the system. Supportive data can then be obtained retrospectively, if the state of the system, for example, software revision and hardware environment, can be determined when the data was produced.

Validation of New Systems During Development

The life cycle approach for new systems is of a general nature and applies to any software development, independent of size, application, hardware, basic software and language used or the nature of the developers (in-house development or vendor supplied). As examples, the life cycle applies to the development of a chromatographic data system as developed by an instrument manufacturer, as well as to the writing by the user of a specific applications program (e.g., a MACRO) as an add-on. The former may have source code of several hundred thousand lines, while the latter may only consist of a few hundred.

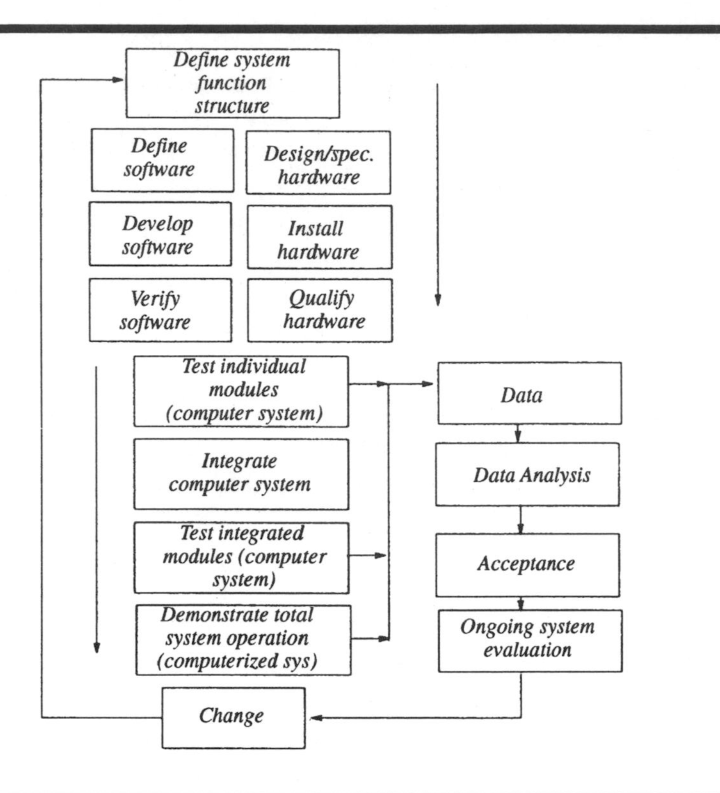

Figure 5.3. PMA's life cycle approach for validation of new computer systems.[7]

A corporate or divisional policy and standards for software development should be available, adequate, and consistently followed and procedures and checklists describing the process should be available at divisional or departmental level. For each project a validation plan should be prepared, reviewed and approved. The plan should describe the procedures that will be used to establish that each component in the system does what it purports to do and that all components perform properly when integrated into a system. Table 5.1 summarizes possible components of a validation plan.

Evaluation of Existing Systems (Retrospective Validation) at the User's Site

Existing computerized systems in laboratories require retrospective evaluation and validation if their initial validation

Table 5.1. Steps for software validation

System scope and functional specifications

Description of the purpose of the system and subsystems or modules and the function they will perform. It shall be written in such a way that it is understood by both the software developer and the user.

Responsibility

The individuals who are responsible for preparing and approving the validation plan must be specifically designated. Responsibilities for the system, its documentation, processes, change control and test arrangements should be defined.

System definition

Defines the test environment with description of hardware, software, communications and other applications that comprise the whole system. Also part of the system definition includes security considerations, special hardware considerations and related documentation.

System design

Describes how defined functional specifications can best be implemented. Alternative solutions are formulated and analyzed and the best solution implemented.

Test plan

As it is difficult to test everything, one should limit the test to the most commonly used functions. Boundaries should be described within which tests should be performed. Limitations of the plan and functions not included in the plan should be documented. The test plan should also include the names of people executing the test. Modifications after review should be documented and authorized.

Test data

The data to be used in the validation plan together with limitations (i.e., when data sets do not cover all possible events) should be specified. The data sets, which can come from previous experiments or studies, should be kept for revalidation.

Test cases and expected results

Test cases and expected results of each test should be listed in the plan, and actual results should be checked against the expected results.

Continued on next page

Continued from previous page

Acceptance criteria

The plan must include the criteria that will be used to formally accept the system.

Revalidation criteria

The plan should include criteria for revalidation of the system after any changes. Depending on the extent of the change, only a partial revalidation may be necessary, but, in any case, a functional test should be done.

Change control

The plan should include a change control procedure valid throughout the life cycle.

Back-up and recovery

The plan should review back-up and recovery procedures for the computer system, providing documented evidence that the computer will continue to do what it purports to do after recovery from a system failure.

Formal review and acceptance of the plan

The plan should be signed off by the system owner and by management and should include a statement that the system is validated.

Archiving

All documentation from the validation procedure shall be archived under fireproof and burglarproof conditions.

---

was not formally documented. It is difficult to use the same validation criteria for an older computer system as those used for a new one. The software might not have been developed in accordance with the most recent product life cycle guidelines, and full documentation may not have been archived.

Fortunately, existing computerized systems have an advantage not shared by new systems: the experience gained over time. The validation process can take advantage of this wealth of historical experience by reviewing the quality of analytical results obtained from computerized systems. Such a review may provide sufficient evidence that the system has done and is still doing what it is supposed to do.

Trill,[67] an inspector from the United Kingdom, reported that systems installed prior to 1989 have generally been poorly documented and controlled. He gave a list of 20 weaknesses that were noted. Agalloco[71] gave guidelines that should be followed for validation of existing computer systems. Kuzel[72] recommended documentation that should be available after validation of existing computer systems. Hambloch[73] gave a strategy for retrospective evaluation with a practical example of a Standard Operating Procedure for retrospective testing. In 1992 Lepore from the US FDA presented a paper at the Sixth International LIMS Conference on "The FDA's Good Laboratory Practice Regulations and Computerized Data Acquisition Systems."[74] In summary, he expects users of such systems to provide evidence to prove that systems operated properly in the past, are currently operating properly and will continue to operate properly in the future.

Before a decision is taken to validate an existing system retrospectively, serious thought should be given as to whether the purchase of or update to a new system might not be more effective. Important criteria to consider are the anticipated costs of a validation versus a new system and an estimation on how successful the validation will be. The latter can be estimated from the history of the computer system, looking for regular maintenance, calibrations and performance checks and trouble free operation over a long period.*

Once the decision has been made to validate the system, a validation plan and documentation should be prepared. Ideally, there should be the same documentation available for existing systems as described in the previous chapter as for new systems, and every attempt should be made to acquire this information. Table 5.2 gives a pragmatic recommendation on the steps to be performed for retrospective evaluation and validation.

The validation protocol for an existing system should include a list of missing documentation usually required for validation. The plan should also provide an explanation as to why

---

***Author's note:** Formal retrospective evaluation and validation of existing systems is a typical example of different perceptions of validation requirements in different countries and industries. While such a need may not be understood and not mandated in an environmental testing laboratory in Europe, if the system has been in use for several years and its performance has been tested through internal daily quality control checks and external proficiency testing, a US FDA investigator may ask for such documentation during a cGMP inspection of a quality control laboratory.

Old versus **New**

Considerations

- **Age?**
- **Availability of documentation?**
- **Uncontrolled changes?**
- **Length of source code?**
- **Language of source code?**
- **Number and types of problems?**
- **Anticipated costs for validation versus new?**

It may be cheaper to upgrade or replace rather than to validate an old system

Figure 5.4. Considerations prior to performing a retrospective validation

Table 5.2. Recommended steps for retrospective evaluation and validation

1. **Describe and define the system**
   - what should the system do
   - operating parameters
   - list of hardware and software with key functions
   - list of accessories such as cables, spare parts, etc.
   - system drawings

2. **Collect any documentation available**
   - reports from internal and external users on number and type of problems
   - validation certificates from vendors for purchased systems
   - internal reports for systems developed in-house (development process, source code, qualification of people, quality assurance principles applied)
   - formulae used for calculations
   - user manuals

3. **Collect information on instrument history**
   - system failure reports
   - maintenance logs and records

Continued on next page

Continued from previous page

- calibration records
- results on performance checks (e.g., results of system suitability tests or quality control charts)

4. **Qualify the system**
   - evaluate information and documentation collected under 2 and 3
   - obtain evidence that the equipment performs as expected, either based on existing tests, or if they are not available, through new tests.

   For system testing a test plan should be developed that is designed to exercise the various functions of the system. Acceptance criteria should be defined before the test starts. After the test phase a formal report that documents the results should be generated. The amount of testing for existing systems can be substantially reduced by relying on operating experience.

5. **Update documentation and develop and implement a plan on how to maintain the performance and security of the system**
   - update system description, specifications, records of operator training, drawings, appropriate SOPs, user manual, safety procedures
   - develop preventive maintenance plan
   - develop calibration schedule
   - develop performance verification schedule
   - develop a system backup and disaster recovery plan
   - develop error recording, reporting and remedial action plan

the documentation is missing. In many cases, the qualification may have been performed but the relevant data not documented. In other cases, the data may have been retained but proper authorization signatures not obtained. The validation plan should also contain a contingency plan that describes what should be done if the previously generated data deemed to be incorrect, e.g., who should be notified.

After the evaluation and validation the following documentation should be available as a minimum:

- The validation plan and protocol
- A description of the system hardware and software with all functions

- Historical logs of hardware with system failure reports, maintenance logs and records and calibration records
- Test data demonstrating that the system does what it purports to do
- SOPs and schedules for preventive maintenance and ongoing performance testing
- Records of operator training
- Error recording, reporting and remedial action plan

Validation of 'Office Programs' Used for Laboratory Applications

Spreadsheet programs are popular in analytical laboratories to evaluate analytical data and to derive characteristics of the analyzed products. Determination of calorific value from a gas analysis is a typical example. Databases are used to correlate data from a single sample analyzed on different instruments and to obtain long-term statistical information for a single sample type. The processes may be automated using MACROs, for example, enabling the analytical data to be transferred, evaluated and reported automatically. In all these programs, analytical data are converted using mathematical formulae. Word processors are used to format analytical data in such a way that they can be presented directly to clients. The question is, should these programs be validated, and if so, to how great an extent?

Today the understanding is that the programs themselves don't have to be validated by the user. For example, no source code inspection is required nor does the user have to obtain evidence of validation from the vendor. It is important to mention that this is the current opinion (1995), which may change in the future.

There should be some documentation on what the program is supposed to do, who defined and entered the formulae and what the formulae are. The user should test and verify the functioning of the program. A frequently asked question is how much testing should be conducted. Testing should

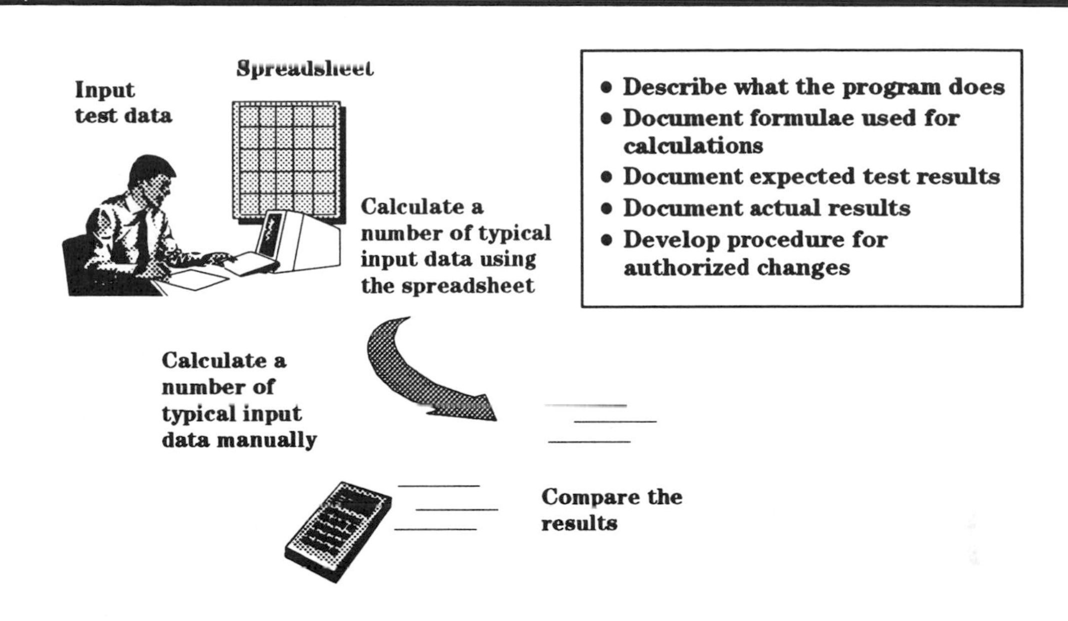

Figure 5.5. Verification of spreadsheet programs and databases

demonstrate that the system is providing accurate, precise and reliable results. A spreadsheet, for example, may be tested using typical examples throughout the anticipated operating range. If the spreadsheet is to be used to further evaluate small numbers, the test program should also include small numbers, or if an accurate calculation requires three digits after the decimal, the test should also use numbers with at least three digits after the decimal. The tests should be repeated thereafter at regular or non-regular intervals.

A big issue in the usage of spreadsheets is the security and the integrity of the spreadsheet, as formulae and MACROs are easily changed. Therefore, procedures should be in place on how to prevent non authorized changes and how to implement authorized changes.

The validation should be documented with the following available as a minimum:

- A description of what the program does with the analytical data.

- A description of the mathematical formulae used in the program.
- A listing of the MACRO program, if any is used.
- A test plan with acceptance criteria.
- Test sheets with anticipated and actual results, signed and reviewed.
- Procedure for changes (e.g., who authorizes and makes changes).
- Date of installation.

Revalidation and Reverification of Software and Computerized Systems

Software has one distinct advantage over hardware: it does not change its performance characteristics over time. Theoretically there should be no need to retest or verify the performance of software as long as the hardware and environmental conditions do not change. However, almost 100 percent of all software written will be changed following its release for use. There are three reasons for a software change:

1. To correct errors
2. To adapt software to changes in its operating environment
3. To enhance the software

When Is Reverification and Revalidation Required?

The performance of computerized systems should be verified periodically and whenever a change is made to the hardware, application software or operating system or when environmental conditions are outside the specification originally made for the computer system. Changes in system software or application software may influence the correct functioning of the computer system. For instance, revalidation should be done if functionality is added to the software. Software developed and validated at the user's site should also be revalidated by the user. Revisions of commercial software should be revalidated by the vendor, who should then provide assurance that the revalidation did indeed take place. Parts of the validation, e.g., acceptance tests, should also be performed after a period of time. Operating procedures should be available that define after what period of time or after what type of changes revalidation or reverification is necessary.

What Should Be Revalidated?

Revalidation does not necessarily mean repeating the entire validation procedure. If the system is well understood, revalidation might mean validating the parts of the system that have been changed plus any related sections affected by the change. The extent of software modification will influence the degree of revalidation. If only a minor cosmetic change to a user interface is made, revalidation may consist of a user interface test, a functional test and of a review of the change control and documents that may be affected. However, one should not underestimate the influence a change in one part of a program can have on other parts, as Clark[37] pointed out: "Software systems, to a greater extent than most other types of systems, can exhibit totally unexpected behavior after they are changed. It is a mistake to assume that a program fix is free of errors. A fix might introduce new errors that have an effect on an area of the system that is far distant from the repaired segment."

For example, if in an HPLC software changes are made in the area of pump control, it is not enough to test only the parameters affecting pump control. In this case the entire HPLC system should go through a functional test. On the other hand, it may not be necessary to verify all details of the report package.

How Should Test Files Be Reused?

When reverification is required, it is recommended to always use the same data sets used during acceptance testing of previous versions of the system. In chromatography, it is wise to use the originally captured chromatographic data files (containing area slices) and to integrate and calculate Area percentage, Norm percentage and External Standard (ESTD) and Internal Standard (ISTD) amounts with the same algorithms used in the first verification. It is strongly recommended to automate this process to avoid human errors and to encourage operators to test more frequently. Automation should also include documentation of test results. Unexpected differences in test results obtained from old and new versions should be identified, documented, resolved and retested.

6. Validation Efforts at the Vendor's Site

In this chapter a practical example is given using Hewlett-Packard's computer-controlled HPLC systems to illustrate the validation process at the vendor's site. Hewlett-Packard (HP) Analytical Products Group (APG) operates all its validation activities according to the specifications defined in the *APG Product Life Cycle* document. Various software development standards were studied and finally consolidated to create the APG product life cycle. The process for the Waldbronn Division, Germany, has been documented by Hans Biesel and is certified by Weinberg, Spelton & Sax, Inc., a company specializing in computer system validation, and by ISO 9001. It was also reviewed by representatives of the pharmaceutical industry in preparation for a US FDA audit. The concept has been published in a primer[76] and in a product note[77] and is explained in more detail in this chapter.

Requirements Analysis and Definition Phase

The software life cycle starts with a requirements analysis and definition phase. It defines the requirements that the product must meet for functionality, usability, performance, reliability, supportability and security. The goal of this phase is to specify both the problem and the constraints upon the solution. Environmental and safety standards are included in the requirements list as well as national and international statutes. Planning activities in this phase include project plans, budgets, schedules and validation, verification and testing. During this phase the project team is established, usually comprising representatives from system development, product marketing, product support, quality assurance,

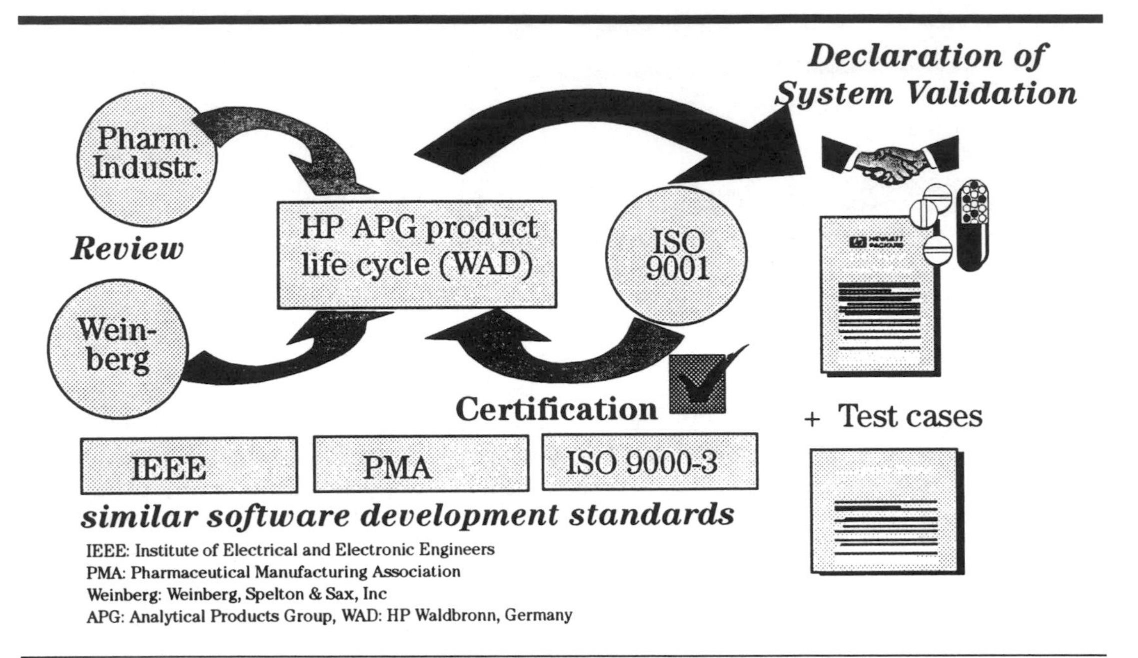

Figure 6.1. Hewlett-Packard's HPLC product development is standardized, has been reviewed by the pharmaceutical industry and independent third party validation companies and is ISO 9001 certified. For each product a declaration of system validation is supplied to users providing proof that the individual product was developed and tested following documented standards.

manufacturing and application chemists, who represent the users and are deeply involved in the development of a functional requirements specification and in the user interface prototyping. A project team leader is appointed to manage the project and a project notebook is created.

Customers from all market segments are interviewed by team members to discover their needs. Activities that cross multiple divisions are coordinated to avoid overlap of software development and to ensure consistency of feature set and user interface across all HP analytical instruments. Finally, a list with all proposed functional requirements is drawn up and evaluated by the project team. Usually the list is too long for all requirements to be implemented within a reasonable time-frame, so the requirements are prioritized into three categories, "Musts", "Wants" and "Nice to haves". "Must" requirements are considered to be those that are a prerequisite to the success of the software and are always included in the final specifications. "Wants" and "Nice to haves" are of lesser importance and are included only if their implementation does not appreciably delay the project.

The external reference specifications (ERS) document is developed that includes an overview of the scope and benefits of the project and a detailed description of the complete product from a user's perspective. The document is reviewed by the project team. It is the starting point for system design and also the basis for functional testing and the user documentation (e.g, user manual, on-line help).

A preliminary marketing plan is developed that describes the target market for the product, addresses the user's needs, does competitive analyses, contains pricing and a sales forecast with product lifetime estimate, contains a plan for product structure, user documentation, training, a market communication (marcom) plan and preliminary strategies for applications development, support and beta tests.

Deliverables for this phase include:

- ☑ External reference specifications (ERS)
- ☑ Risk assessment
- ☑ Marketing plan
- ☑ Quality plan

Design Phase

The goal here is to design a solution that satisfies 100 percent of the defined requirements and falls within the set constraints. Alternative solutions are formulated and analyzed, and the best are selected. Verification activities during the design phase include inspecting the product for completeness, correctness, consistency and checking that the design directly correlates with the defined requirements. Thorough design inspections are of utmost importance because correction of defects detected in this phase is much less costly than when detected in a later life cycle phase.

Details on system screen designs, report layouts, data descriptions, system configurations, system security, file design, system limitations and memory requirements are laid out by system developers and are usually formally inspected with members of the development team. A major output of this phase are the internal design documents and prototypes. The design documents are based on the ERS and can be used as a source for the technical support documentation.

In this phase operators from different backgrounds test the user interface in a process called user-interface prototyping to determine the access to the intended function and their understanding of interaction concepts.

Deliverables for this phase include:

- ☑ Internal design documents
- ☑ Reports on design inspections
- ☑ Usability test report update
- ☑ GLP validation/documentation plan
- ☑ Chemical performance and application test plan
- ☑ QA plan update
- ☑ Alpha test plan

Implementation Phase

In the implementation phase the detailed design is implemented in source code, following written and approved software coding standards, and results in a program that is ready to be tested. After certain groups of functions have been programmed, they are tested individually by the programmers before they are integrated into a larger unit or into the complete system. Verification includes code and internal documentation, test designs and all activities that determine whether the specified requirements have been met. Concurrently with the implementation phase, system documentation, such as user manuals, is prepared. Documentation also includes a description of the algorithms used by the program.

In this phase the system also undergoes a rigorous usability evaluation with testers from different backgrounds. The goal is for an experienced chromatographer to be able to perform the basic functions without the need for formal instruction (the so-called plug-and-play approach).

Deliverables for this phase include:

- ☑ Source code with documentation
- ☑ Code inspection/walkthrough reports
- ☑ Documentation of test cases in preparation for the test phase
- ☑ Marketing plan update

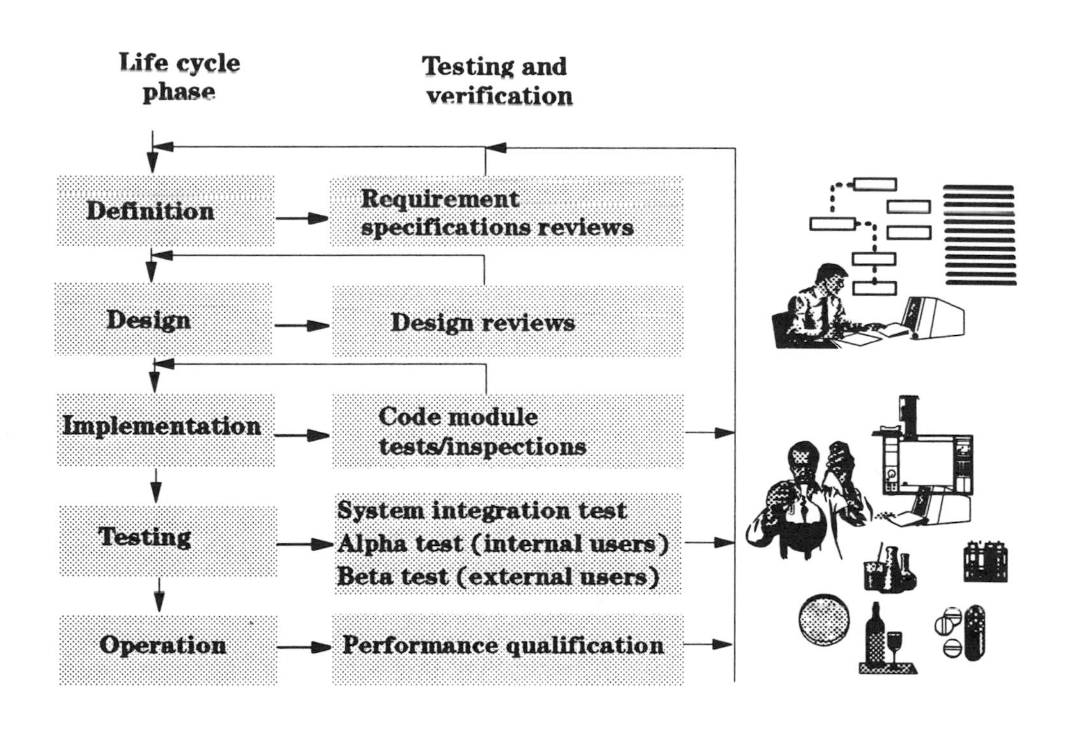

Figure 6.2. Testing and verification are done throughout all life cycle phases

Test Phase

Thorough testing and verification are most important for any validation. For a software project testing and verification are done throughout all life cycle phases. The goal is to detect errors as early as possible, if there are any. Requirements specifications and the design are reviewed or inspected during the definition and design phases and the code is tested and may be formally inspected by the programmers during code implementation. Proper functioning of software together with the equipment hardware is verified in the test phase and during operation.

Types of Testing

Software testing can be classified as being either structural (white box) or functional (black box). Structural testing (white box) of software is the detailed examination of the internal structure of code, including low- and high-level path analysis and inspection for adherence to software development procedures. It tests logical paths through the software by providing test cases that exercise specific sets of conditions. Besides the source code, other documentation, such as logic diagrams, branch flow analysis reports, description of modules, definition of all variables and specifications of all inputs and outputs, are required.

Functional testing (black box) of software evaluates the outputs of a program compared to the expected output values for a range of input values. For a computer-controlled analytical system, functional testing should always include analytical hardware to verify proper parameter communication and data flow. Source code is not required, but a full set of system specifications and a description of functional routines, such as calibration algorithms, must be available.

Structural testing is done in the development department and starts during the implementation phase. Code modules are checked individually by the programmers and may be formally inspected by peers, if appropriate. Modules are then linked together into a system and tested as a whole for proper functionality to ensure that designs are correctly implemented and the specified requirements satisfied.

Written in advance, the test plan defines all test procedures with their pass/fail criteria, expected test results, test tasks, test environment for equipment and computers, criteria or acceptance and release to manufacturing and the persons responsible for conducting these tests. The test plan also specifies those functions excluded from testing, if any. Individual tasks cover functional testing, simulation of incomplete functions as integration proceeds, mathematical proof of results, records of discrepancies, classification of defects and corrective actions.

Alpha-Testing

After the programmers have completed the first round of corrections, the system undergoes functional testing in typical operating conditions, so-called alpha-testing. Over several weeks, groups of chemists and other professionals conduct the testing, using test cases defined for each person in a test book that must be signed off by individuals on completion of the test. The objective is to test the complete computerized system for functionality, usability, reliability, performance and supportability as stated in the ERS. The user manual is prepared before the alpha-test to allow test personnel to verify its accuracy and usefulness. At least one test case requires installation of the software and hardware according to the installation instructions.

The system is not only tested under typical operating conditions but also at the limits under which it will be required to operate—an approach known variously as *worst case* testing, *gray box* testing or testing of *boundary conditions*. Testing boundary conditions is important because most software errors occur around its boundary limits. Combinations of several worst cases are also tested. For example, if the spectral data acquisition rate of a UV-visible diode-array detector is specified as 10 spectra/second, the wavelength limits as 190 to 600 nm and the number of UV diode-array detectors connected to a single computer as two, the system is tested to perform well under the combination of worst cases, that is spectral acquisition over a range from 190 to 600 nm, at a data rate of 10 Hz from two diode-array detectors. Software testing also includes so-called stress testing. Inputs with inappropriate character types, alphabetic characters instead of numeric ones, for example, or inappropriate character length and character composition are made, and instrument parameters that lie outside the instrument's operational limits are entered. The expectation is that these inputs will not damage data or disrupt system and software operation and that the system will recover after producing error messages.

Beta-Testing

Once software defects and usability discrepancies reported during alpha-testing have been corrected, the software may be tested at selected customers' sites (the so-called beta-test). The key feature of beta-testing is that it is conducted in a customer environment and supervised by a person not involved in the development process. One of the objectives of beta-testing is to test the HP product delivery and support channel. A trained HP applications engineer (AE) assists the customer with installation and checks the software installation procedure.

Defect Tracking and Response System

Documenting software errors is important and problems should not be casually reported for repair by the programmer on an *ad hoc* basis. Problems found during testing are tracked using the HP internal Defect Control System (DCS). After product release all errors are tracked by HP's Software Tracking and Response System (STARS). Defects, classified as low, medium, serious and critical defect density and defect summaries are recorded for access throughout the organization. The number of critical defects after the last test cycle must be zero for the software to pass the release criteria. Finally, to ensure that the software is ready to be installed, plans are revised as necessary and all other coding, testing and documentation is completed. Once released, a system is in place to notify customers should any problems arise.

Deliverables for the test phase include:

☑ Test plans with acceptance criteria and test cases

☑ Test results

☑ Validation documents

☑ Defect density report

☑ User training material

Release for Production and Installation

After the testing is complete and the code is corrected for errors, the software is released for production and distribution. The product is considered ready for release when it has met all the criteria specified in the quality plan and after formal sign-off by product line, quality assurance and manufacturing management. Prerequisite for this is sufficient training of service engineers who must be able not only to install and operate the software but also to train users and answer users' questions. Availability of user documentation in the form of on-line help and printed reference material is also a release criterion.

The manufacturing department ships the product in accordance with manufacturing guidelines, based on the receipt of valid purchase orders. The product documentation includes a *Declaration of System Validation* with statements from Hewlett-Packard that the software was developed and tested according to the Hewlett-Packard Analytical Software Life Cycle, a process that has been certified for ISO 9001 quality standard compliance and that has been reviewed by representatives from both computer validation companies and the pharmaceutical industry.

In preparation for installation at the user's site the user receives information on space, power and environmental requirements. The user prepares the site by ensuring the correct environment as specified by the supplier and may install special power or wiring for the hardware. During installation, computer hardware and any interfaces to the analytical equipment are installed. All wiring is checked for proper connection. Operating and application software are loaded, and peripherals such as printers and analytical instruments are configured. The cabling and electrical functions of the complete system are tested in the operating environment.

The installation also includes a familiarization training where operators are instructed on how to use the system. A customer representative should formally sign off a form indicating that the installation has been completed and that the system is accepted.

Operation and Maintenance

Once the product is released, customer problems and enhancement requests are received and tracked through normal HP support channels. Defects reported by users are recorded, classified and corrected in updated versions. Enhancement requests are evaluated by an expert team consisting of users of the system, and marketing and R&D engineers who make a proposal to management on software changes. Changes to the software will always require a partial or full revalidation. A formal change control system addresses changes to validated systems and includes quality assurance review/approval.

Hewlett-Packard's software contract service keeps customers informed on new and updated revisions. Those customers who did not purchase the contract service may still receive

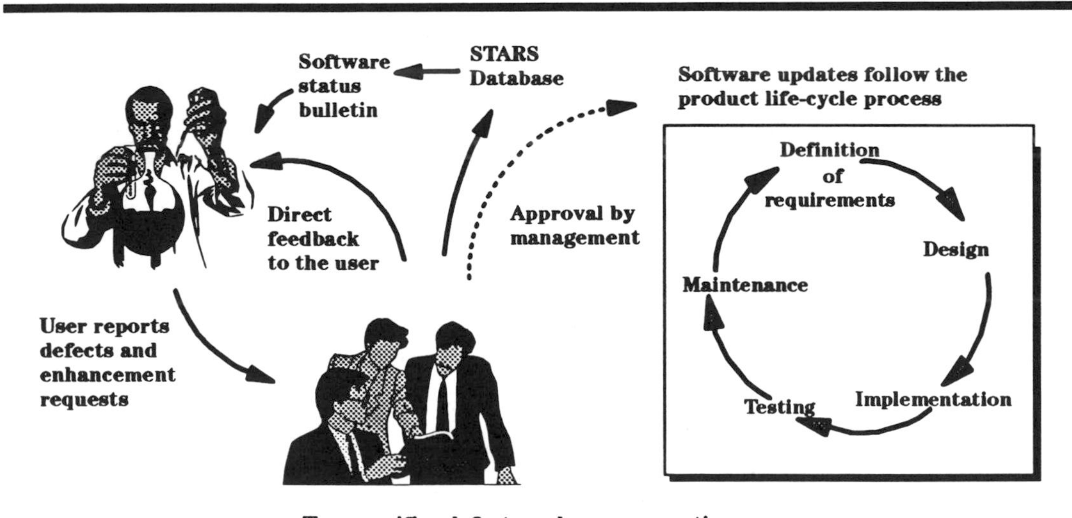

Figure 6.3. Handling of defect reports and enhancement requests

information on new and updated software through *PEAK,* HP's journal for analytical laboratories.

Change Control

A thorough *change control* procedure is very important in the overall validation process and maintains a tight control over why, how, when and by whom the system was changed and who authorized the changes. Reported software defects and functional enhancements requested by users and accepted as valid are examined for their impact on verification activities. Any changes to the hardware or software of a computerized system must be clearly specified as if it were a new system. A control system determines the degree of revalidation required according to the extent of the changes to the requirements list, design specification and implementation plan. The impact on the entire system of changes within one module and the necessity for revalidation or reverification are evaluated by a software development committee. Principally, software redevelopment and testing follow the same procedure as for newly developed software. Compared to a new product the amount of testing and verification is reduced through intensive use of regression testing.

All details of the changes are thoroughly recorded and documented, together with details of completed tests and their results. Software status bulletins describing any known defects and giving recommendations on temporary work-arounds accompany all software revisions.

The software documentation also includes a revision history that clearly describes the improvements made from one revision to the next, for example, defects repaired and new features. The impact that the changes may have on application software running on previous versions is also listed. For example, if a chromatographic data station can be customized using MACROs, the revision documentation provides a list of all MACRO commands that have either changed or no longer exist. A specific set of software is characterized by a product name or number and a revision number. It is an absolute must that only one unique source code be associated with such a characterized software.

Documentation

The documentation of all activities during software development, testing and maintenance is important because it may be required by the user for an FDA audit. Documentation of software is also important to facilitate communication among those who develop, market, manufacture and finally support the systems; and for those software development engineers who implement changes, good documentation is a must.

The source code for Hewlett-Packard products is well documented. Development environment, directory structure and compiling procedure, code and comment lines, and data flow charts are collated and stored in a disaster resistant archive on a third party site. Like other life cycle documents, it is accessible at Hewlett-Packard for review by authorized regulatory auditors.

The entire system documentation is fully developed during software development and not as an afterthought once the system is running. Validation documentation is available to users who perform the tasks required for proper installation, operation and performance qualification and includes:

- Functional specifications
- Test cases and expected output
- Procedures for ongoing performance verification

Further documentation is available on signing of a non-disclosure agreement:

- Test results as obtained during factory validation
- Description of software development and validation process

Algorithms and formulae are documented in a user manual entitled *Understanding Your ChemStation.* This is important to verify correct functioning in the user's environment.

7. Responsibilities of Vendors and Users

Software may be developed in-house, purchased from a software development company or purchased and then customized in-house. With the publication of the *Blue Book* in 1983,[3] the FDA expressed its concerns about software quality. One of the FDA's chief concerns has been that much of the software used by regulated pharmaceutical firms was developed in the non-regulated vendor sector.[8] Thus, when an organization plans to purchase software instead of developing its own, that organization is responsible for the overall validity of the software. This is stated in the OECD GLP principles (see Figure 7.1).[50] This does not mean that the user is left alone to deal with the problems. The OECD principles also make a statement that validation can be done at the vendor's site on behalf of the user. This is also practice of the US FDA as it was spelled out very clearly by Tetzlaff, a former FDA investigator: "The responsibility for demonstrating that systems have been validated lies with the user."[35]

Without prior written assurances from the vendor, the laboratory should not automatically assume that vendor supplied software has been properly validated: "User companies should not assume, without documented evidence, that vendor supplied software has been properly validated. This assumption is frequently unfounded and FDA investigators continue to find firms using automated systems without a documented basis to ensure proper performance."[35]

The US Pharmaceutical Manufacturer Association's Computer System Validation Committee published a paper on *Vendor-User Relationship*,[78] defining vendors as external or internal and stating that both should be treated alike. The paper recommends the establishment of a vendor assessment

- **It is the *responsibility of the user* to ensure that the software programme has been validated.**
- **It is acceptable for formal validation of software to be *carried out by the supplier on behalf of the user*, provided that the user undertakes a formal *acceptance test*.**
- **The user should ensure that all software obtained has been provided by a recognized supplier. *ISO 9001 registration is considered to be useful*.**

OECD GLP Principles, Monograph 49

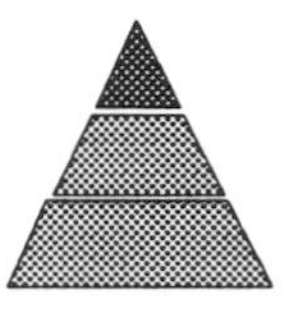

Figure 7.1. Vendor/user responsibilities according to OECD GLP Principles[50]

program that may include formal written procedures for the selection, evaluation and auditing of vendors. The paper also emphasizes the user's responsibility for regulatory compliance. "Although a strong partnership between the user and vendor is a desirable goal, the manufacturer's (user's) regulatory responsibilities must always be borne in mind."

Software Categories

As described earlier, software can be divided into three different categories:

1. System software, e.g., operating systems such as UNIX® or Microsoft® DOS.
2. Canned standard applications software, for example, off-the-shelf chromatography software, generally supplied by an instrument vendor.
3. User specific applications software, written by the user to customize the system.

To date there has been little discussion amongst computer validation experts on how system software should be validated. No one has suggested that a user needs to own or inspect the source code for an IBM®, Microsoft®, or DEC® operating system.[18] It is assumed that every time an applications program is executed, its operating system is exercised as well. Even when the program has been on the market for only a short time, its wide and frequent use is assurance of the program's success—programs with serious bugs simply do not survive in the marketplace. Further, it is reasonable to expect software development companies wishing to remain competitive to supply system software that uses quality standards and procedures in its development.

Standard chromatography software is usually developed by an instrument or software vendor. Part of the validation is done at the vendor's site but the user should still perform functional tests and ongoing performance checks in his or her environment. In some cases software with the same functionality may be developed at different sites. For example, chromatography software may include a statistical package for the evaluation of the instrument's precision. In such cases the vendor usually validates the complete system, while the user tests operation and performance. If the vendor's software does not include such a package, the user may write a program to do statistics using special application program interfaces (commands) supplied with the vendor's software. In this particular case the vendor will not take any responsibility for the accuracy and reliability of the user-contributed program and the user should validate this software. Often instrument purchasers share such contributed programs in the form of MACROs. Here, the user must take full responsibility for validation and testing even when the MACRO in question was distributed by the vendor.

Validation is relatively easy if the customized application software is written as add-on MACROs to an existing software, because the vendor supplied source code remains unchanged. A different scenario exists if customization was achieved by making changes to the source code. Here the situation is more complicated: the user must access the source code and only the user can assure proper functioning of the complete software. Such software can be validated by the purchaser directly or by business consultants active in this area.

A vendor may, at the request of and in close cooperation with the user, develop application software as an addition to the standard software. A typical example is the development of a special HP ChemStation MACRO in a local regional office. In this case there should be a clear list of requirement specifications signed by the vendor and the user. The users should either validate the software themselves, or should obtain assurance from the vendor that the software development and validation was followed and documented according to the vendor's development and test procedure. In this case the users should also perform the functional testing in their environment.

Occasionally a laboratory may commission the development of a software package for a user specific application from a specialist vendor. In this case the user must get assurance that the software was developed and validated according to documented software development standards. The user may have to audit the developer's QA department to obtain this assurance. Alternatively, the vendor can provide evidence of its conformation to reliable software development standards through audits by third parties, e.g., a recognized computer validation firm, a pharmaceutical company or an ISO 9001 registrar.

In this section we give practical advice on the interaction between vendors and purchasers, in particular when the user's firm purchases a standard software package not specifically written for the user.

Validation Responsibilities

Validation activities for software purchased from a vendor are generally shared between the vendor and the user. Specific guidelines on the vendor-user relationship on purchased software can be found in the UK Pharmaceutical Industry Supplier Guide: *Validation of Automated Systems in Pharmaceutical Manufacture.*[39] It is recommended that user requirements, functional and hardware specifications be made by a team consisting of users and vendors (see Figure 7.2). For standard software, e.g. chromatography software, the vendor collects inputs from a representative mix of anticipated users to make

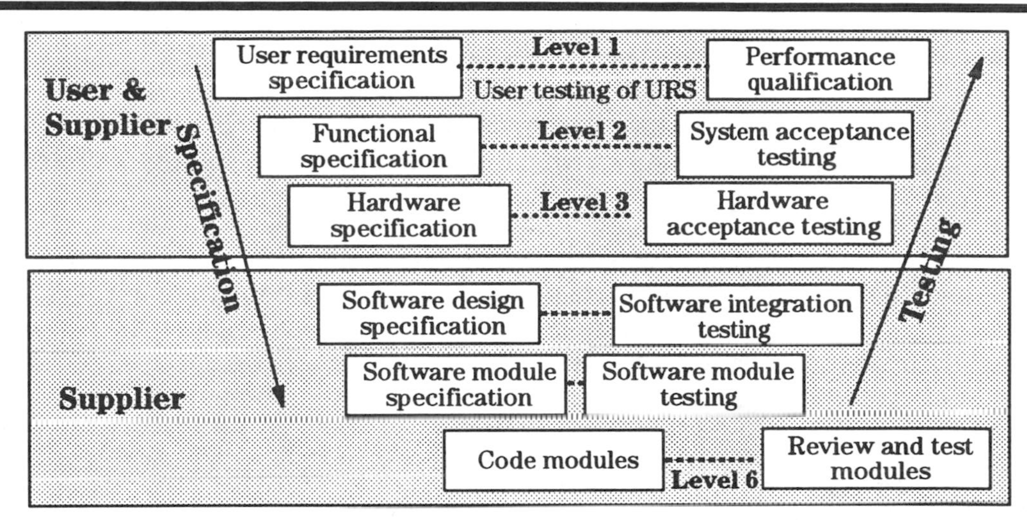

Ref.: UK Pharmaceutical Industry Supplier Guidance: *Validation of Automated Systems in Pharmaceutical Manufacture*, available from LOGICA, U.K. tel. +44716379111

Figure 7.2. The vendor/user responsibilities for system validation are described in the UK Pharmaceutical Supplier Guidance (draft): *Validation of Automated Systems in Pharmaceutical Manufacture.*[41] For computerized analytical systems the vendor drafts user requirements specifications and functional specifications based on inputs from a representative mix of anticipated users.

sure that the product will meet the requirements of target users. Typically, the vendor drafts a set of user requirement specifications based on these inputs as part of the overall development.

Hardware, as well as system acceptance and performance testing, should be done at both sites while software design, implementation, source code review and software module and system integration testing should be done by the vendor.

Vendors must take as much care in developing software as user firms themselves would take. Certain validation considerations can be addressed at the contract or purchase-agreement stage. For example, a firm may request documented assurance that acceptable software development standards have been followed by qualified individuals during the development of the software. These conditions may be assured either by an audit of the vendor's firm by the user's firm or by a third party. Here the advantage of ISO 9001 registration becomes obvious. Companies registered for this quality standard should have software development processes in place that follow internationally accepted quality assurance standards for software development, for example, ISO 9000-3.

A declaration on successful development validation should be supplied with each software, giving additional confidence that the product purchased has passed the company's validation procedure. It is also recommended that the supplier guarantees accessibility of validation documents in case a regulatory body requires them for inspection.

Source Code Availability

The source code is needed in order to alter any software program (e.g., to correct malfunctions). If the user does not change the source code of a commercially available software, a structural line-by-line source code testing is not be required of the user. To sit down and browse through the source code looking for errors is virtually worthless. It has been estimated that to conduct such low level line-by-line inspection would require 40 man-years for 1.5 million lines of code,[41] assuming that the team of experts works no more than two one hour sessions daily and that one person can examine 100 to 150 lines per day. The effort involved in such a source code inspection would be of the same magnitude as writing a new software; therefore, source code for non-user specific software purchased from a vendor, such as operating systems, statistical packages and GC control and evaluation software, generally need not be available to the user (as per recommendation by PMA).[7]

Even though the official FDA *Policy Guide 7132a.15* states that the drug manufacturer must maintain source code for vendor-supplied application software, this policy is rarely enforced in practice. It is however strongly recommended that users obtain assurance from vendors that the source code will be made accessible for inspection by regulatory agencies should the need arise. "To further protect the user, contract agreements might stipulate that archival history and source code of all relevant versions will be deposited in a third party vault that is accessible to the user in case the software vendor goes out of business."[7]

Vendors generally react positively to this. Hewlett-Packard for example, officially states in its *Declaration of System Validation* which is shipped with each HPLC, CE, UV-visible and GC ChemStation, that the source code for these products can be inspected by regulatory agencies. For products developed at the Waldbronn site in Germany, the source code is locked in a bank safe. Sisk[79] reported that Perkin-Elmer had formalized its methodology for escrow accounts, placing copies of PE Nelson's software product, source code and engineering records with an independent third party. If the company should go out of business or no longer support the particular product, the customer has the right to obtain source code documentation in order to continue to use, change or develop the product as necessary. Chapman[80] gave an overview on source code availability and vendor-user relationships. The article includes proposals for typical contemporary contract agreements between vendors and users.

Vendor Testing and Specifications

Along with their software, vendors should supply a list of specifications with the relevant environmental boundary conditions. They should also provide a list of the computer system tests performed prior to the product's manufacturing release to establish environmental product specifications including temperature, humidity, radio frequency and other electromagnetic interferences. Frequently these tests are performed in compliance with national or international standards, such as CSA (applicable in Canada and the United States) and VDE (applicable in Germany).

Software should go through rigorous performance testing at the vendor's site during and on completion of development. It is the FDA's position that test data should be provided to the user, even though some users find this to be impractical: "Some users of vendor supplied software have taken the position that it is not practical or realistic to expect vendors to provide validation test data or supporting evidence. The FDA

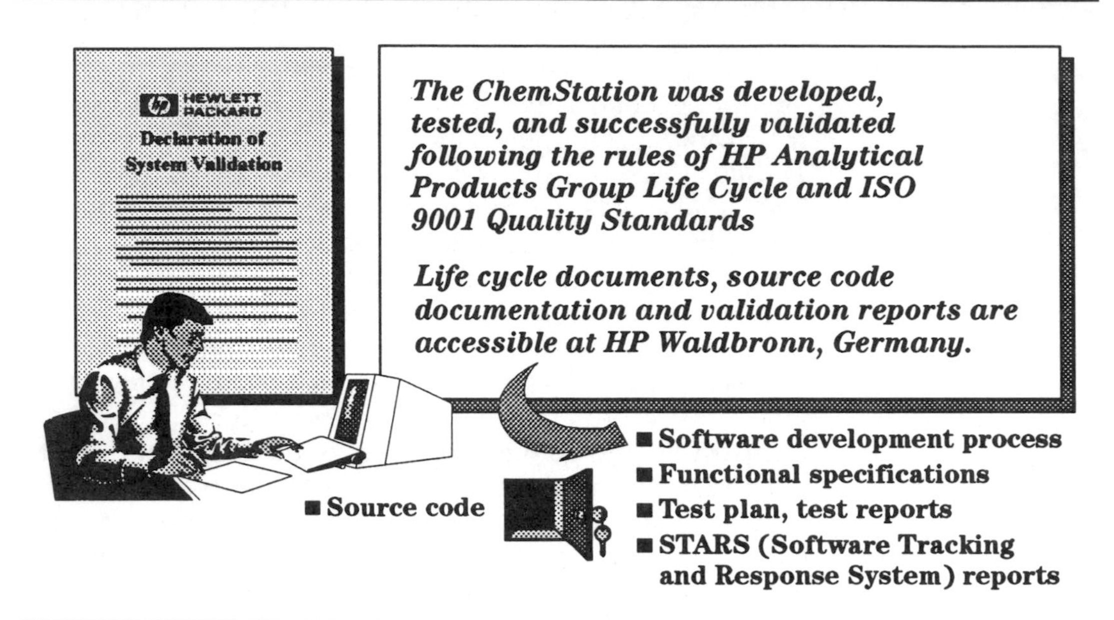

Figure 7.3. Hewlett-Packard validates its HPLC ChemStation products following the HP Analytical Product Life Cycle (WAD) and ISO 9001 Quality Standard. A declaration of conformity to these standards is supplied with each product. The source code is locked in the safe of a bank and can be made accessible to regulatory agencies.

considers this to be an unrealistic position. Companies should consider alternative vendors when they encounter suppliers who are unable or unwilling to share test data or evidence to support system performance."[35] Also the OECD GLP consensus document *Compliance of Laboratory Suppliers with GLP Principles* makes this quite clear on page 7: "Suppliers are expected to provide all information necessary for the correct performance of the instrument."[50]

As Clark[37] pointed out, "The validation documentation that is provided by the vendor does not need to be voluminous. It must be sufficient to allow the pharmaceutical manufacturer to conclude that the validation performed by the vendor was thorough and accurate. The vendor should retain all validation data and make it available to the user on request. It is inadequate for the vendor to simply certify that the system was validated."

User Testing

The instrument should be tested on site before it is used for the first time, within certain time intervals and after repairs. "All computer software, including that obtained from external suppliers, should normally be acceptance tested before being put into service by a laboratory"[50] and "It is the responsibility of the user to undertake an acceptance test before use of the software programme. The acceptance test should be fully documented."[50] During this acceptance testing, proper functioning of the software for the intended application should be verified. This type of test is frequently referred to as *functional testing* or *black box testing* because source code is not needed. For an HPLC ChemStation with software for instrument control and data evaluation this means checking the LC instruments for correct response as well as testing that peak integration, quantification and printout function correctly. If the ChemStation software includes acquisition and evaluation of spectra, typical spectral evaluation routines, e.g., peak purity checks and peak confirmation with spectral libraries should be tested.

The vendor should provide assistance either directly or through provision of test packages to aid reverification of the system according to its specifications. Test files and procedures (preferably automated) should be available to the user to reverify the computer system in situ with a few keystrokes. In chromatography these consist of chromatograms, integration and calibration routines and routines for automated comparison of results stored on a disk with the newly calculated results. Documented procedures that describe how to verify the performance of an automated system should be supplied as part of the user instruction kit.

Support

The vendor should offer fast, high quality support throughout the lifetime of the system, ensuring selection of the correct configuration at purchase and optimal performance throughout installation and operation. The vendor should also provide comprehensive training to the user on how to operate the system. Phone support should be available to help solve software problems and operational difficulties while prompt on-site support should be readily available for preventive maintenance and in case of hardware problems.

Combined Standard and User Contributed Software

Frequently, complex computer systems will run a combination of in-house developed and purchased software. A simple example is the customization of a ChemStation by the user using MACRO programs to further automate the generation of data. As pointed out earlier, validation of the add-on software should be done by the user. Minimum validation documents should include:

- ☑ overview statement of the purpose of the software
- ☑ user requirement specification
- ☑ description of the software
- ☑ listing of the program
- ☑ test plan with expected results
- ☑ tests carried out with test result (Calculate the result manually on a few typical examples.)
- ☑ user's guide with installation procedure
- ☑ training documentation
- ☑ guidelines for periodic performance verification
- ☑ security procedures for changes
- ☑ name of the author
- ☑ approvals
- ☑ date of installation

The development of an automated test procedure, which can be executed as often as possible, is recommended. The tests should be performed with a set of data that generates known results.

- **Overview statement: purpose**
- **Requirement specification**
- **Program description, flow charts and printouts**
- **Test protocols and results**
- **User's guide**
- **Training documentation**
- **Security procedures for changes**
- **Name of author and approvals**
- **Date of installation**

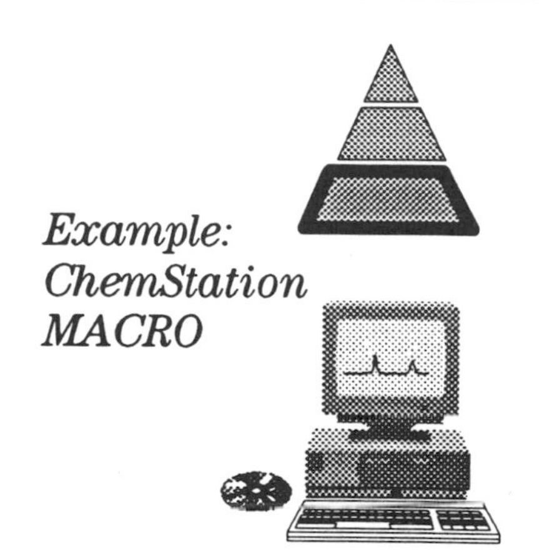

Chemists tend to change and update MACROS frequently without sufficient documentation.

Figure 7.4. Documentation that should be available after validation of user specific application programs, for example, HP ChemStation MACROS

For such systems vendor updated software revisions may be critical, especially if the updated version will have an effect on the interface between the vendor's and the user's software (e.g., if the meaning of MACRO commands has been changed). The user should get information from the vendor on how the updated version may affect the interface. The user should test his or her software after it has been integrated into the vendor's updated standard software.

A similar validation procedure should be applied to programs written by laboratory personnel to perform calculations or generate reports with a special format. Such programs should be validated, the effort required depending on the size and complexity of the program. If the program consist of a few lines only, it may be inappropriate to go through all of the steps that are required for large programs. In this case it may be sufficient to:

1. describe the software and what it is supposed to do

2. develop the software

3. develop a test plan with expected results
4. calculate the result manually on a few typical examples
5. calculate the results using the software and compare the results with the manually calculated results
6. write a procedure describing the management and security for the program
7. document everything

Step 5 could be automated and performed periodically, for instance every time the program is used to generate results for an FDA report. If the results are always documented, the system is basically reverify each time it is used.

The same concept can be applied to spreadsheet programs or databases, such as Lotus 1-2-3® or MS-Excel™. In this case it is not necessary to validate Excel or Lotus themselves, but, especially when the results are obtained through MACROs, some test data should be routinely verified for accuracy and the results documented.

Typical responsibilities of vendors and users are summarized in Table 7.1.

Table 7.1. Summary table with typical tasks for vendors and users

Vendor	User
• Have a proper quality system in place for software development and maintenance	• Ultimate validation responsibility
• Design	• Define user requirement specifications for a specific project
• Structural testing	• Select and qualify the vendor
• Functional testing	• Select suitable instrument and options

Continued on next page

Continued from previous page

Vendor	User
• Provide specifications	• Acceptance testing (functional testing in the user's environment)
• Generate and maintain validation documentation	• Routine maintenance
• Provide user documentation	• Inform vendor of failures and propose enhancements
• Provide environmental specifications and operational limits	• Validate analytical methods
• Provide mathematical equations and indicate where used	• Perform ongoing system suitability tests
• Provide protocols for acceptance testing	• Keeps records of maintenance
• Provide operator training and support	• Perform periodic reviews
• Provide validation documents	• Ensure system security
• Establish software tracking and response system	• Develop appropriate SOPs
• Archive source code preferably in third party vaults	• Ensure adequate operator qualification
• Ensure accessibility of source code to regulatory agencies	
• Provide tools for system security	

8. Calibration, Verification and Validation of Equipment

Compared to software and computer systems, the validation and verification of instrument hardware is relatively easy. Analytical hardware is frequently one module in a computerized system and typically less complex. Specifications and test conditions are available, and errors can be identified during its installation and initial operation. Hardware does, however, have one disadvantage when compared to software; with time some parts may deteriorate or become contaminated. This can influence the performance and reliability of an analytical system. An HPLC detector flowcell, for example, can become contaminated with the result that less light is transmitted. This will have an effect on the detector's baseline noise, thereby increasing the method's detection limit.

The focus of hardware validation is more on maintenance, cleaning and performance verification and covers those parts and characteristics that may change their performance as a function of time. Physical deterioration over time should be monitored, for example, degrading lamp intensity in a UV detector. Because this will affect the performance of the instrument, frequent performance checks of the detector are required.

As with software and computer systems, validation and verification of analytical hardware is done at both the vendor's and the user's sites.

Analytical hardware validation activities accompany the equipment through the entire product life cycle. A validation model has been proposed in the UK Pharmaceutical Industry Supplier Guidance: *Validation of Automated Systems in Pharma-*

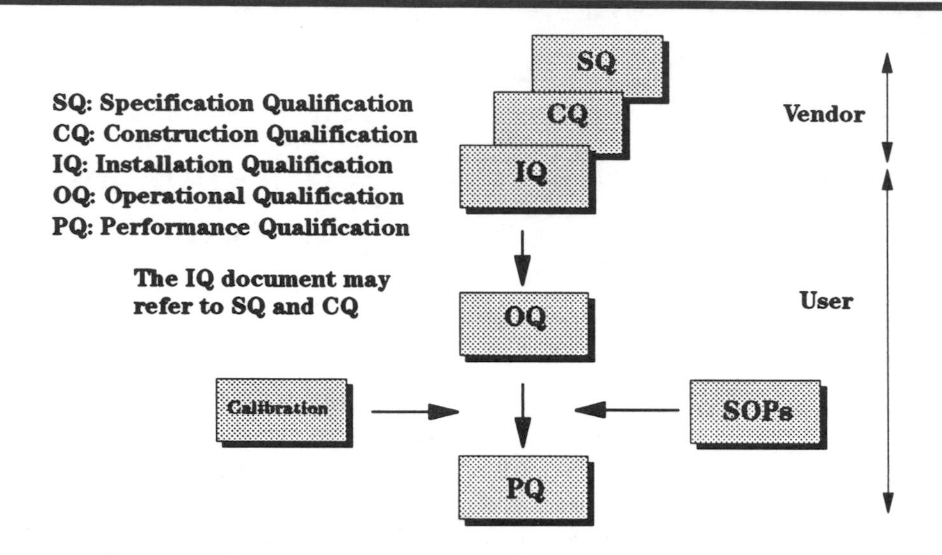

Figure 8.1. The equipment validation model[41] and responsibilities of vendors and users of commercial off-the-shelf (COTS) products.

ceutical Manufacture.[41] The life cycle is divided into different phases: specification, construction, installation and operation. Formal checks are done during each phase to ensure that the product complies with the user requirements and functional specification.

Usually analytical equipment is purchased from an outside company. Validation activities occur at both the supplier's and the user's sites. While the supplier is responsible for proper definition, design, development, manufacturing and testing, the user should verify the performance in his or her own environment and should have an ongoing maintenance and performance control system in place to ensure a high quality of data during operation.

All product development, manufacturing and testing activities should follow well documented procedures as part of a quality management system. The quality management system should be evaluated and certified by an independent third party following national or international quality standards, for example, ISO 9001.

Validation/Verification at the Vendor's Site

As an example the development and validation process as practiced by Hewlett-Packard Waldbronn Division in Germany is briefly described. The product development and validation process for hardware follows similar principles as described in more detail for software in chapter six of this book. As with software validation, the development process is divided into different phases with formal checkpoint meetings after each phase. Checkpoint approvals depend on successful review and acceptance of predefined project deliverables.

Any product development activity starts with an investigation and evaluation of users' requirements, collected by a combination of direct visits, phone interviews and mail questionnaires. The external reference specifications (ERS) or functional specifications are derived from this investigation. These requirements not only include performance characteristics, such as detection limits and linearity, but also items such as usability, reliability, serviceability, purchasing price and cost of ownership.

A project team is formed comprising members of the quality assurance, development, marketing, manufacturing and administration departments. Product development is divided into different phases where various prototypes from breadboard through laboratory prototype to manufacturing prototype are built. At a relatively early stage in the product development it is important to verify the compliance of the product to the functional specification, including checks on usability, reliability and serviceability. Throughout all phases, the project's financial performance is evaluated to ensure that the projected selling price and the company's financial goals are congruent.

Test procedures are developed to characterize the performance of the instrument at each stage of its development, on installation at the user's site and during ongoing operation.

To estimate the annual failure rate and to verify that the instrument can meet all its performance specifications, long-term performance tests are made under normal and extreme environmental conditions. For example, instruments are placed in an oven and operated at elevated temperatures and high humidity to simulate tropical conditions. Results of these tests are also used to predict the annual failure rate of the instrument.

To ensure a product's safety radio frequency interference tests and other tests are performed in compliance with national or international standards, such as CSA (applicable in Canada and the United States) and VDE (applicable in Germany).

User documentation is written during the development phase including operating manuals on how to use, optimize and maintain the instrument, a service manual on how to repair the instrument, and a performance test manual on how to verify the instrument's performance. If necessary, this documentation is written in a step-by-step instruction form such that it can be used and archived as formal Standard Operating Procedures (SOPs).

Before being released to manufacturing, the instrument goes through extensive functional and performance testing, performed by technicians or chemists in both the company's applications laboratory and externally in the users' laboratories. During this test, the instruments are used for as many applications as possible to ensure applicability to a wide range of possible samples.

A procedure is developed to ensure that each instrument meets its functional design and performance specification when it is shipped from the factory. This may be a performance test for each major specification on each instrument or statistically based tests for less critical specifications. A *Declaration of Conformity* according to the European Norm EN 45014[81] and ISO/IEC Guide 22[82] is shipped with all instruments to document that the instrument operated within specification when it was shipped from the factory. The *Declaration of Conformity* is an extract from detailed and comprehensive test documentation and includes the following information:

- The name and the address of the supplier
- Clear identification of the product (name, type and model number)

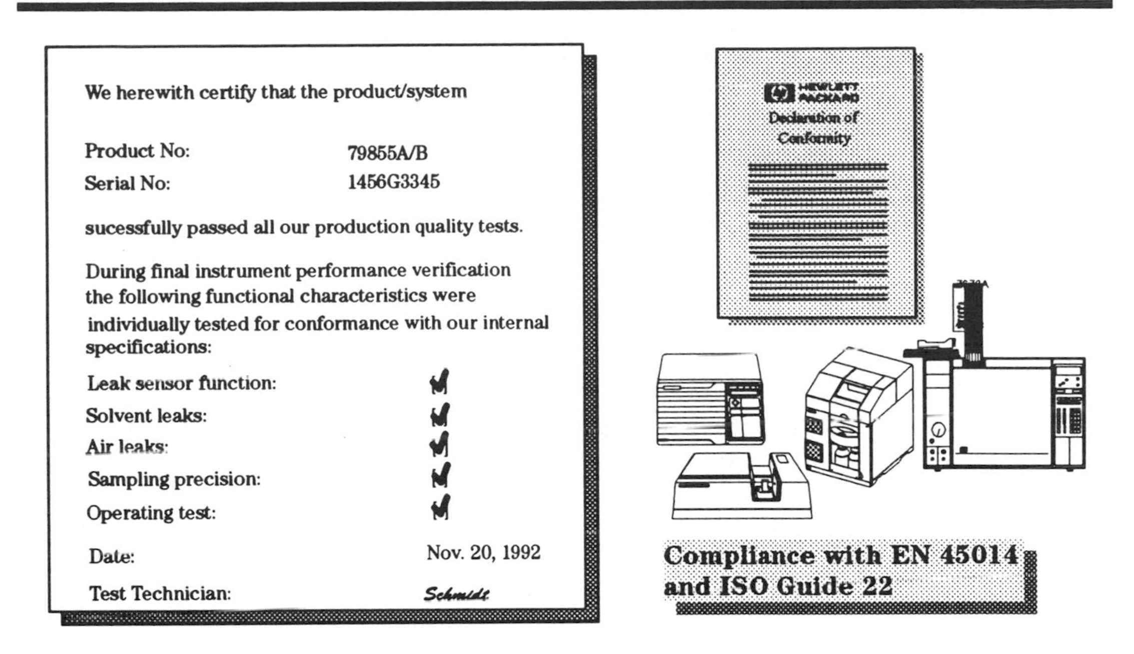
We herewith certify that the product/system

Product No: 79855A/B
Serial No: 1456G3345

sucessfully passed all our production quality tests.

During final instrument performance verification the following functional characteristics were individually tested for conformance with our internal specifications:

Leak sensor function: ✓
Solvent leaks: ✓
Air leaks: ✓
Sampling precision: ✓
Operating test: ✓

Date: Nov. 20, 1992
Test Technician: Schmidt

Figure 8.2. A declaration of conformity according to EN 45014 and ISO/IEC Guide 22 is shipped with each instrument and documents compliance to specifications.

- Place and date the declaration was issued
- Name and signature of the supplier's authorized person
- Listing of tested items and check boxes with pass/fail information

Once the instrument is shipped to customers, any failure reported by users is recorded in a database. Each failure is analyzed and associated to a specific manufacturing process or a specific part. The database also includes a list of the parts that fail most frequently. If necessary, the quality assurance department initiates a corrective action plan that is implemented by the development and manufacturing departments.

Activities at the User's Site

A prevailing assumption amongst laboratory personnel is that all instruments obtained from reputable manufacturers and shipped via reputable carriers have already been completely tested and that the instruments will meet published specifications when they arrive at their destination. This is a mistaken assumption. Frequently manufacturers only test a limited set of specifications for each individual instrument or test only a certain percentage of shipped instruments for adherence to specifications.

There is a second reason why users should verify the performance; the environment may be different. For example, electric motors used close by the instrument or connected to the same electrical supply may generate enough electrical noise to disturb the signal transfer between a detector and a computer system.

Even before the instrument arrives, users can contribute to ensuring a high ongoing quality standard from the equipment. The user's site activities can be divided into three phases:

- Purchase
- Installation
- Operation

The main activities of each phase are listed in Table 8.1.

The selection of the right instrument configuration should be made by an expert team consisting of users and vendors. On the user's side experts from the laboratory, the quality assurance unit and the company's computer integration department should be involved. The latter is important when wishing to integrate the hardware into a laboratory's data handling and networking system.

For more complex systems a vendor's technical expert should be available to ensure selection of an optimal configuration for the user's application and to make certain that all options and spare parts required for proper functioning of the system are included. A contract between the user's and the vendor's firms should be signed that includes items on the warranty period, coverage of warranty (parts, labor, on-site, off-site), support, preventive maintenance, definition of standard

Table 8.1. Main activities for equipment users

Purchase

- Select the correct analytical technique.
- Select an appropriate vendor.
- Select an instrument with the right capabilities and specifications.

Installation

- Prepare adequate rooms, space, and electrical, gas and other supplies according to the vendor's environmental specifications.
- Check instrument shipment for completeness, for example, cables, documentation, all listed accessories and options.
- Make functional test.
- Train operator.

Operation

- Verify performance throughout anticipated operating ranges.
- Develop and establish an ongoing performance verification program.
- Develop and establish a preventive maintenance program.

installation and any additional installation tasks, for example, demonstration of the instrument for a specific application.

In the following chapters users' site activities are discussed in more detail.

9. Selection and Qualification of a Vendor

The most important step when considering validation of computer-controlled analytical systems used in a regulated environment is the selection of an appropriate vendor. Without the vendor's help and documents it is impossible to demonstrate that the computerized system has been validated during its development and the vendor's assistance can speed up the validation tasks usually done by the user, for example, acceptance testing and operational qualification. In this chapter we give some guidelines on how to select and qualify a vendor and provide criteria for equipment selection.

Selecting a Vendor

At a GLP convention held in Germany in 1991, one of the speakers presented a list of recommended criteria for the selection of instrument vendors in regulated studies. Interestingly, the most important criteria quoted were not instrument performance related but rather focused on a vendor's knowledge of GLP and quality standard issues, their support services and provision of validated software, verified hardware and packages for on-site validation.

How Well Do They Know Good Practice Regulations and Quality Standards?

This is the first and most important question. If the company is not familiar with GLP and quality standards, they cannot provide products and services that help users to meet them.

Will They Allow an Audit?

There may be a need to audit suppliers of software and computer systems, either directly or through third parties. Suppliers who refuse to be audited should not be considered for future purchases.

Do They Provide Adequate Long-Term Support?

You should ask questions about response time, proper training of staff and what happens after product obsolescence. Large companies can generally give more reliable long-term support because they are better established in the market place.

Do They Build All the Required Functions into the Product?

Equipment used in regulated and quality standard environments should be regularly tested, and errors should be identified and recorded. Built-in functionality designed to meet this requirement helps to reduce laboratory costs.

Do They Supply Source Code to Regulatory Agencies?

Regulatory agencies want assurance that source code is accessible to them should they wish to inspect it. Source code is also required for complete software validation, which includes structural testing. The vendor should give written assurance to prospective users that the source code will be made accessible at the request of regulatory agencies. The source code should be located in a safe place outside the vendor's building and be accessible to the user should the vendor go bankrupt.

Do They Supply Details of Algorithms?

Testing of software at the user's site normally includes verification of results, for example, the calculation of amounts from peak areas in chromatography. Mathematical algorithms required to do this verification should be included in the user documentation.

Do They Have an Adequate Quality System?

There is a growing trend in industry towards dealing only with companies whose quality systems are compliant to internationally recognized quality standards, for example, ISO 9001. For vendors not compliant to such a standard, proof should be provided that a similar quality system with documented procedures for product design, development, manufacturing, testing, distribution and servicing is followed. The quality system should ensure that the products meet the design specification when shipped to the user.

Will They Assist with On-Site Validation?

Equipment should be tested, verified and qualified at the user's site at installation, after updates, after repair and after extended use. The vendor should provide validation packages, for example, test procedures, to reduce costs for the user.

Vendor Qualification

Regulatory agencies and the pharmaceutical industry recommend users of standard applications software used in pharmaceutical manufacturing to qualify the suppliers either by obtaining references or by performing vendor audits directly or through third parties.

Vendor audits are performed by the pharmaceutical industry as part of their overall validation plans. Hewlett-Packard, for example, has had several of these audits. Sisk[79] from Perkin-Elmer reported that PE Nelson's Cupertino facility undergoes at least one customer audit per month. If such audits were to be done in all contractual situations, however, both suppliers and customers would expend a large amount of their resources. It was recommended at the *Management Forum Conference*[41] that customers can share their experience with suppliers if the audit items are identical. "If another customer has already audited the supplier for the same reason, then subject to that customer agreeing to share that information, another audit may not be necessary".[39] This is the case for standard software that does not contain customer-specific applications, Hewlett-Packard's ChemStation software, for example. Details on the items for and conducting of an audit are covered in Appendix C of the above document. It is recommended that the audit should be prepared, executed and documented in accordance with guidelines laid down in ISO 10011 Part 1.[75] Even though ISO 9001 is not considered to have the level of technical depth required by the FDA and the pharmaceutical industry,[42] the use of this or similar quality systems will facilitate the audit. Documentation such as operating procedures, test plans and test results is already in place and readily available.

A good recommendation is to send the vendor a checklist with the questions typically asked during an audit. If all questions are answered satisfactorily an audit may not be necessary. Table 9.1 provides a checklist of items that can be used when auditing suppliers. Information in this table comes from the author's experience with several audits, the UK Pharmaceutical Industry Supplier Guidance and from literature references. For example, the US PMA's Computer System Validation Committee with Christoff and Sakers[78] published a paper on vendor-user relationships that includes recommendations for vendor assessment. Abel[83] also gave recommendations for conducting audits.

Table 9.1. Checklist for supplier assessments

Audit Item
Company information
☑ Company history: how long has the company been in business?
☑ Financial status (obtain copy of annual report)?
☑ Is the company currently in the process of negotiation for sale?
☑ Size (number of employees?)
☑ What percentage of sales was/is invested in R&D?
☑ Does the vendor have an established customer base?
☑ Are there company policies on quality, security etc.?
☑ Is there a Quality Management System?
☑ Is the vendor compliant to ISO 9001? (obtain copies of certificates)
☑ Has the company been audited by other companies?

Continued on next page

Continued from previous page

Organization

- ☑ Is there a formal quality assurance department?

Software development

- ☑ Does the vendor follow engineering standards (e.g., ISO 9000-3, IEEE, ANSI)?
- ☑ Is there a software quality assurance program?
- ☑ Is there a structured methodology (e.g., life cycle approach)
- ☑ Are there life cycle checkpoint forms (checklists)?
- ☑ Is all development done at the vendor's site?
- ☑ If not, are third parties certified or regularly audited by the vendor (e.g., ISO 9001)?

Testing

- ☑ Who develops test plans?
- ☑ Are requirement specifications reviews and design inspections done periodically?
- ☑ How is functional testing performed?
- ☑ Who is testing (outside the development department)?
- ☑ Are there test protocols?
- ☑ Are there procedures for recording, storing and auditing test results?
- ☑ Who approves test plans/protocols?

Support/training

- ☑ How many support personnel does the company have?
- ☑ Does the vendor have formalized training programs in installation, operation and maintenance of systems?
- ☑ Which support systems are in place (phone, direct)?
- ☑ Where is the nearest office for support?
- ☑ Is a service contract available and what does it cover (installation, startup, performance verification, maintenance, training)?
- ☑ Does the company provide consulting and validation services?
- ☑ Do support people speak local language?
- ☑ What is the average response time?
- ☑ Is the service organization compliant to an international quality standard (for example, for ISO 9001 or ISO 9002)?
- ☑ How long are previous versions supported and at what level?
- ☑ How long is support and supply of parts guaranteed?
- ☑ Is training available on how to use the system? Location, frequency?
- ☑ Are training materials available (description, media)?

Continued on next page

Continued from previous page

Failure reports/enhancement requests

- ☑ Is there a formal problem reporting and feedback system in place?
- ☑ How are defects and enhancement requests handled?
- ☑ How are customers informed on failure handling?
- ☑ Are quality records existing?

Change control

- ☑ Who initiates changes?
- ☑ Who authorizes changes?
- ☑ Are there procedures for change control?
- ☑ Do they include impact analysis, test plans?
- ☑ Is there a formal revision control procedure?
- ☑ Will all updates get new version numbers?
- ☑ Are there procedures for user documentation updates?
- ☑ How are customers informed on changes?

People qualification

- ☑ Do people have knowledge on regulatory compliance and programming science?
- ☑ Is there documentation on education, experience, training?

The product/project

- ☑ When did development of the software first begin?
- ☑ When was the initial version of the software first released?
- ☑ How many systems are installed?
- ☑ How often are software releases typically issued?
- ☑ How many employees are working on the project?
- ☑ Are there functional specifications?
- ☑ Are there samples of reports?
- ☑ Which vintage of data files can be processed with today's software

User documentation

- ☑ Are there procedures and standards for the development and maintenance of user documentation?
- ☑ What documentation is supplied?
- ☑ For how long is the documentation supplied?

Archiving of software and documentation

- ☑ What is archived, for how long (software, revisions, source code, documentation)?
- ☑ Where is the source code archived?

Continued on next page

Continued from previous page

- ☑ Can the source code be made accessible to regulatory agencies?
- ☑ Is a periodic check of data integrity ensured?

Security

- ☑ Is the developer's area secure?
- ☑ What type of security is provided to prevent unauthorized changes?
- ☑ Are there written procedures specifying who has access to software development and control?

Equipment hardware

- ☑ Is there a formal and documented equipment life cycle standard?
- ☑ Is there a product specification conformance verification system?
- ☑ Are functional specifications available?
- ☑ Are environmental specifications available?
- ☑ Are test reports of individual products stored and for how long?
- ☑ Are calibration procedures/tools validated and traceable to national or international standards?
- ☑ Are techniques used to establish product quality?
- ☑ Is there a performance history regarding on-time shipping?
- ☑ Are there documented installation procedures?
- ☑ Do they provide operating procedures for performance verification at the user's site?
- ☑ Do they provide services for performance verification services at the user's site?

Customer training

- ☑ Does a training manual exist?
- ☑ Do they provide tools for training, e.g., computer-based or video training?
- ☑ Do they offer operator training courses (frequency, language)?
- ☑ Is there documented evidence for the trainers' qualifications?

10. Installation and Operation

Three steps are required to put a computerized system into routine operation:

1. preparation of the site for installation

2. installation of hardware and software

3. operational, acceptance and performance testing

It is important to perform both operational and acceptance testing in the user's environment even if individual modules or the complete system were previously tested at the vendor's location. The performance and suitability of the complete system should be verified for its intended use before and during routine operation.

Preparing for Installation

Before an instrument is delivered, serious thought must be given to location and space requirements. A full specification for bench or floor space requirements, environmental conditions, such as humidity and temperature, and utilities, such as electricity, compressed gases and water, should be obtained from the vendor well in advance. Care should be taken that all environmental conditions and electrical grounding are within the vendor's specified limits and that the correct cable types are used. All safety precautions and recommendations, for example, the location of radioactive measuring devices or devices generating electromagnetic fields, should be considered and acted upon.

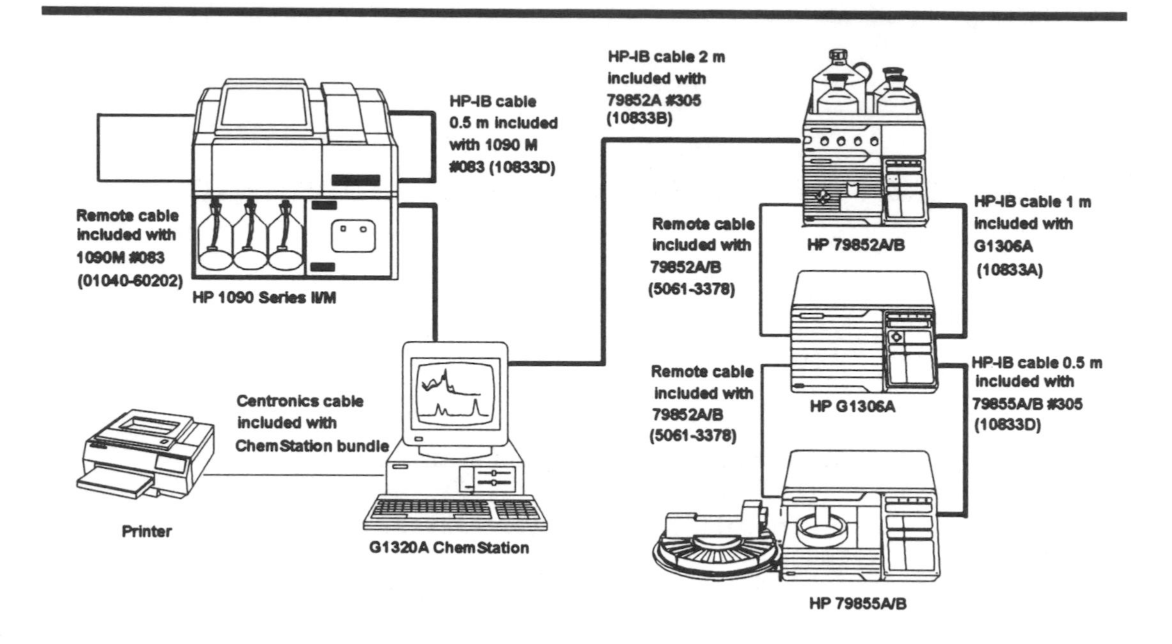

Figure 10.1. Example for hardware configuration of an automated HPLC system. A single computer is used to control two diode-array detector based HPLC systems. IEEE (HP-IB) interfaces are used to ensure fast data transmission.[84]

Installation

On arrival of the system the user should check that the shipment is complete and confirm that the equipment received matches the original order. A visual inspection of the entire hardware to detect any physical damage should follow, and other items such as cables, accessories and documentation should be checked. After all cables and tubings are connected, an electrical test of all the modules and the system should follow. The impact of electromagnetic fields generated by electrical devices near to the computer should be considered and evaluated. For example, electromagnetic energy emitted by a nearby fluorescent lamp or by a poorly shielded electric motor can interfere with the minute voltages transmitted by sensors and detectors. Recommendations on how to check for different types of electrical noise are found in a paper published by Alford and Cline.[11] For more complex

Table 10.1. Form for computer system identification

Computer hardware	
Manufacturer	
Model	
Serial number	
Processor	
Co-processor	
Memory (RAM)	
Graphics adapter	
Video memory	
Mouse	
Hard disk	
Installed drives	
Space requirement	
Printer	
Manufacturer	
Model	
Serial number	
Space requirement	
Operating software	
Operating system (version)	
User interface (version)	
Application software 1	
Description	
Manufacturer/vendor	
Product number (version)	
Required disk space	

Continued on next page

Continued from previous page

Application software 2	
Description	
Manufacturer/vendor	
Product number (version)	
Required disk space	

instrumentation wiring diagrams should be generated, if these were not obtained from the vendor.

The installation should end with the generation and sign-off of the installation report referred to in pharmaceutical manufacturing as the Installation Qualification (IQ) document. IQ is a term that is defined by the US PMA's Computer System Validation Committee as: "Documented verification that all key aspects of hardware installation adhere to appropriate codes and approved design intentions and that the recommendations of the manufacturer have been suitably considered."[12] It is recommended that documented procedures with checklists for installation and preprinted forms for the installation report be used.

On completion of the installation procedure, both hardware and software should be well documented with model, serial and revision numbers.

For bigger laboratories with large amounts of equipment a computer database for storage of instrument records is preferable. Entries for each instrument should include:

- ☐ In-house identification number
- ☐ Name of the item of equipment
- ☐ The manufacturer's name, address, phone number for service calls and service contract number, if applicable
- ☐ Serial number and firmware revision number of equipment
- ☐ Software with product and revision number
- ☐ Date received

- ☐ Date placed in service
- ☐ Current location
- ☐ Size, weight
- ☐ Condition when received, for example, new, used, reconditioned
- ☐ List with authorized users and responsible person

It is recommended that all important documentation be photocopied, with one copy placed close to the instrument and the other kept in a safe place. A sticker should be affixed to the instrument with information on the instrument's serial number and the company asset number.

Logbook

A bound logbook should be prepared for each instrument in which operators and service technicians record all equipment related activities in chronological order. Information in the logbook may include:

- ☐ Logbook identification (number, valid time range)
- ☐ Instrument identification (manufacturer, model name/number, serial number, firmware revision, date received, service contact)
- ☐ Column entry fields for dates, times and events, for example, initial installation and calibration, updates, column changes, errors, repairs, performance tests, quality control checks, cleaning and maintenance plus fields for the name and signature of the technician making the entry

Table 10.2 includes a summary list on documentation that should be existing or generated for equipment before and during installation.

Operator Training

Adequately trained operators are an important requirement for the proper functioning of an analytical system because the

Table 10.2. Documentation that should be received from the vendor or generated before and during installation and updated during operation

Operating Manuals

Manuals are usually supplied by the vendor. They are mainly used during the start-up phase of the instruments and later on as a reference. They should include a list of spare parts with ordering information. Manuals for software and computer systems should include algorithms used in the program.

SOPs

SOPs can be supplied by the vendor or can be developed in-house. They should be used for day-to-day operation of the systems and should be available for:

- Basic operation
- Routine maintenance
- Calibration
- Performance testing

Logbook

This should be a bound book with serialized page numbers. Templates can be supplied by the vendor. Entries should be made by the user and should include:

- Logbook identification (number, valid time range)
- Instrument identification
- Records of installation with initial calibration, operational and performance tests
- Ongoing records for
  —maintenance
  —calibrations
  —performance tests
  —events (e.g., failures and repairs)

Binder

The binder should include documents supporting logbook records

- Instrument brochure (if there is one)
- Instrument specifications (supplied by the vendor)
- Declaration of conformity to specifications for equipment (supplied by the vendor)
- Declaration of system (development) validation for software and computer systems (supplied by the vendor)
- Lists with intended use, anticipated functions and operational range (developed by the user)
- Installation protocol

Continued on next page

Continued from previous page

- Installation, operational and performance qualification and requalification protocols (for GMP-regulated laboratories)
- Additional documentation for calibrations and performance tests, for example, instrument plots and print outs, certificates, evidence of personnel qualified to do calibration, etc.
- Additional information on repairs, for example, protocols and service reports

best computerized system cannot produce consistent and correct results if the operators make mistakes. To ensure that operators are familiar with all the tasks they are expected to perform they should receive training on instrument operation, parameter optimization and system maintenance. They should also be able to recognize any system malfunction and report it to the appropriate person, for example, the laboratory supervisor. An ongoing program to familiarize users with the system should be implemented, and the effectiveness of the training measured and documented.

Preparing for Operation

Further activities may be required to ensure that equipment is functioning correctly before being used routinely. For example, the fluid system of a high performance liquid chromatograph requires purging with a mobile phase to elute any gas bubbles or organic residues and if the instrument is to be used for gradient operation, it should also be purged under gradient conditions. A gas chromatograph used for trace analysis may require passivation before operation to avoid any adsorption of the trace compounds.

Operation

After installation of the hardware and software an operational test should follow, a process referred to in pharmaceutical manufacturing as Operational Qualification (OQ). OQ has been defined as "The process of demonstrating that the equipment will perform consistently as specified over all intended ranges."[41] The US PMA's Computer System Validation Committee defined OQ as: "Documented verification that the system or subsystem performs as intended throughout representative or anticipated operating ranges."[12] The goal is to demonstrate that the system operates *as intended* in the user's environment.

For an HPLC system operational testing may include, for example, verification of correct communication between modules, checking of baseline noise and precision of retention times and peak areas. Vendors should provide operating procedures for these tests that include acceptance criteria and recommendations in the event criteria cannot be met.

The words *intended ranges* in the definition of operational qualification are important. They mean that those instrument functions and limits that will not be used do not require testing. This also means that the user should specify the *intended range* before the operational qualification testing begins. For example, an HPLC pump with four channel proportioning valve designed to perform gradient analysis need only be tested for isocratic operation if the pump is used for isocratic runs only. Or, if an HPLC UV-visible detector will always be used to measure relatively high concentrations, it is not necessary to measure its performance close to its detection limit.

This also has an impact on the maintenance efforts required. To keep the baseline noise of a UV-visible detector within the manufacturer's specification, frequent cleaning of the flow cell, ultrapure mobile phases and frequent lamp changes are required. For applications not requiring the highest sensitivity, this effort may be reduced by raising the acceptance limit for baseline noise.

For systems consisting of multiple hardware modules, an HPLC system, for example, operational testing may be performed for the entire system or for each module. Testing of the complete system is referred to as holistic testing, while the testing of individual modules in a computerized analytical system is known as modular testing.[85] For this discussion it should be noted that testing of parameters like flow rate precision or injection volume precision in practice requires more than one HPLC module. For example, testing the precision of an autosampler always requires a pumping system and a detector, while testing a pump's precision by injecting a series of standards and measuring the standard deviation of the peaks always needs an injection system and a detector. Some important characteristics are influenced by several parts of the system, for example, peak area precision. This is influenced by the repeatability of the injection volume and by the stability of the solvent delivery system. In some instances, especially when using standards that produce peaks with signal-to-noise ratios below 100, the integration repeatability may determine the system precision. Because of these interdependencies, and because proper functioning of all individual modules does not necessarily mean that the system as a whole meets its specification, holistic testing is preferred. For example, the baseline noise of an HPLC refractive index detector is not only influenced by the detector characteristics, but also by a smooth pump flow. A detector with relatively poor performance may produce reasonable results with an excellent pump, but may be unacceptably noisy when connected to an average pump.

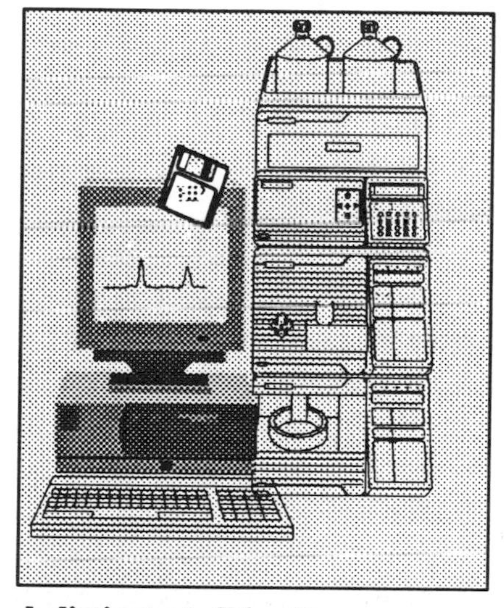

holistic: test all in all (recommended for performance testing)

modular: test module by module (recommended for some calibrations, for example, wavelength accuracy, and for diagnostics)

Figure 10.2. Holistic versus modular testing

It is always recommended that a system test be performed. Modular testing is recommended for troubleshooting purposes, where individual modules may be interchanged to identify problem sources, and for calibrations, such as UV-visible detector wavelength accuracy, which can be performed independently of other modules.

The installation and operational testing of equipment may be performed by either a vendor's representative, by a third party or by the user. In either case the installation and testing should follow written protocols and the manufacturer's recommendations.

Before and during routine use the instrument should undergo performance testing that demonstrates its capability to perform as expected under real-life conditions, a process referred to in pharmaceutical manufacturing as Performance Qualification (PQ). PQ has been defined as, "The documented verification that the process and/or the total process-related system performs as intended throughout all anticipated ranges."[41] For a computerized analytical system this means demonstrating that a specific system works as expected with a particular method, better known in analytical laboratories as system suitability testing.

It is also recommended that the need for method validation or revalidation on the new equipment is evaluated. This may be necessary when newly developed methods are used or when the new equipment was not covered during method validation.

Table 10.3. Steps for routine use of computerized systems

Activities	Documentation
Before Installation	
• Obtain manufacturer's recommendations for installation site requirements. • Check the site for compliance with the manufacturer's recommendations (space, environmental conditions,	• Manufacturer's recommended site preparation document • Checklist for site preparation

Continued on next page

Continued from previous page

Activities	Documentation
utilities such as electricity, water and gases). • Allow sufficient shelf space for SOPs, operating manuals, disks.	
Installation (Installation Qualification)	
• Compare equipment received with purchase order (including software, accessories, spare parts and documentation). • Check equipment for any damage. • Install hardware (computer, equipment, fittings and tubing for fluid connections, columns in HPLC and GC, power, data flow and instrument control cables). • Install software on computer following the manufacturer's recommendations. • Make back-up copy of software. • Configure peripherals, e.g. printers and HPLC modules. • Evaluate electrical shielding (are there electromagnetic field sources nearby?). • Prepare installation reports. • Train operators.	• Copy of original purchase order • Equipment, accessory and documentation checklists • System schematics and wiring diagrams • Record of operator training • Installation Protocol
Preoperation	
Examples • Flush HPLC fluid path with mobile phase. • Passivate gas chromatograph if necessary. • Calibrate wavelength accuracy of HPLC UV-visible detectors.	• Notebook and/or logbook entries on calibration

Continued on next page

Continued from previous page

Activities	Documentation
Operation (Acceptance Testing, Operational Qualification)	
• Document anticipated functions and operational ranges of modules and systems. • Perform basic application software functions, for example, integration, calibration and reporting using data files supplied on disk. • Perform basic instrument control functions from both the computer and from the instrument's keyboard, e.g., switch on the detector lamp and the pump and set different wave-lengths and flow rates. • Test the proper operation of equip-ment hardware for anticipated functions. • Document all the operational tests. • Sign the installation and operational protocol.	• Specifications on intended use and anticipated operating ranges • Test procedures of the computer system with acceptance criteria limits and templates with entry fields for instrument serial number, test results, corrective actions in case criteria are not met and name and signature of the test engineer • Procedures with templates for operational testing of equipment describing the test details and acceptance criteria • Signed installation and opera-tional protocols. If the installa-tion is performed by a vendor, the protocol should be signed by both user and vendor representatives
After Installation	
• Affix a sticker to the instrument with the user firm's asset number, manu-facturer's model and serial numbers and firmware revision. • Identify and make a list of all hardware. • Make a list of all software installed on the computer. • Develop a procedure and a schedule for ongoing preventative mainten-ance, calibration and performance verification. • Prepare and maintain a logbook with entries of any instrument problems.	• Instrument sticker • Equipment identification forms (in-house identification, name and model, manufacturer, serial number, firmware and software revision, location, date of installation) • List of software programs with software revisions, disk storage requirements and installation dates • Procedures and schedules for preventive maintenance,

Continued on next page

Continued from previous page

Activities	Documentation
• Develop a computer system security strategy (if not already existing within the company).	calibration and performance verification
Ongoing Performance Control (Performance Qualification, System Suitability Testing, Analytical Quality Control) for Routine Analysis	
• Combine instrumentation with analytical methods, columns and reference material into an analysis system suitable for running the unknown samples. • Define system suitability specifications and acceptance limits for the above system. • Define type and frequency of system suitability tests and/or analytical quality control (AQC) checks. • Perform the tests described above and document results. • Develop SOPs with definition of raw data and for verification of raw and processed data (if not already existing within the company).	• Test protocols • Data sheets with acceptance criteria for system performance and test results • Quality control charts • SOPs for definition of raw data and verification of processed data

11. Validation of Analytical Methods

Analytical methods used to generate data in GLP studies or for use in the quality control of drug manufacturing should be validated and the resulting data included as part of the package submitted to regulatory agencies for new drug applications. Method validation is equally important to those laboratories seeking accreditation for EN 45001,[58] NAMAS[61] and ISO/IEC Guide 25[59] and to everybody concerned about the quality of their data.

Method validation is the process used to establish that the performance characteristics of an analytical method meet specifications that relate to the intended use of the analytical results. Methods need to be validated before their introduction into routine use and whenever the method is changed. To obtain the most accurate results, all of the method variables should be considered, including sampling procedure, sample preparation, chromatographic or electrophoretic separation, detection, data evaluation and the matrix of the intended samples. The proposed procedure should go through a rigorous validation process. The validity of an analytical method can only be verified by laboratory studies, and all validation experiments used to make claims or conclusions about a method's validity should be documented in a report.

A laboratory applying a specific method should have documentary evidence that the method has been appropriately validated. "The responsibility remains firmly with the user to ensure that the validation documented in the method is sufficiently complete to meet his or her needs."[60] This holds for standard methods, for example, from ASTM, ISO or USP, as well as for methods developed in-house. If standard methods are used, it should be verified that the scope and validation data, for example, sample matrix, linear range and detection limits comply with the laboratory's analyses requirements; otherwise, the validation of the standard method should be repeated using the laboratory's own criteria. Full validation of a standard method is also recommended where no information on type and results of validation can be found in the standard method documentation. All validation results should be documented.

Strategies for Method Validation

Validation of analytical methods should follow a well-documented procedure, beginning with the definition of the scope of the method and its validation criteria and including the compounds and matrices, desired detection and quantitation limits and any other important performance criteria. The scope of the method should include the different equipment and locations where the method will be run. For example, if the method is to be run on one specific instrument in one specific laboratory, there is no need to use instruments from other vendors or to include other laboratories in the validation experiments.

Before an instrument is used to validate a method, its performance should be verified (using standards). For example, if detection limit is critical for a specific method, the instrument's specification for baseline noise and for some detectors also the response to specified compounds should be verified. Operators should be adequately trained and/or experienced in the use of the instrument, and any material used to determine critical validation parameters should be checked. During method validation the parameters, acceptance limits and frequency of ongoing system suitability tests should be defined and Operation Procedures for method execution and ongoing performance testing should be developed.

Parameters for Method Validation

The parameters of what constitutes a validated chromatographic method have received considerable attention in

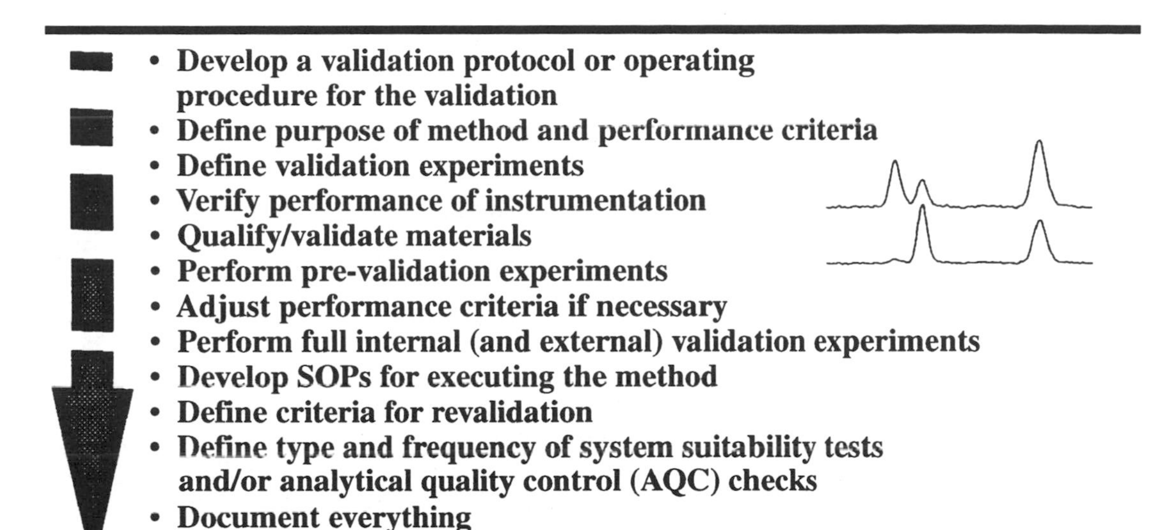

Figure 11.1. Strategy for method validation

the literature and from regulatory agencies.[68,86–99] The *Guidance on the Interpretation of the EN 45000 Series of Standards and ISO/IEC Guide 25* includes a chapter on the validation of methods[60] with a list of nine validation parameters. The International Conference on Harmonisation (ICH) of Technical Requirements for the Registration of Pharmaceuticals for Human Use[93] has developed a draft consensus text on the validation of analytical procedures. The document includes definitions for eight validation characteristics. The United States Pharmacopoeia has published specific guidelines for method validation for compound evaluation.[68] There are no official guidelines referring to biological fluids. The pharmaceutical industry uses methodology recently published in the literature.[94,100–101] The most comprehensive document was published as the Conference Report of the Washington Conference on *Analytical Methods Validation: Bioavailability Bioequivalence and Pharmacokinetic Studies* held in 1990 (sponsored by the American Association of Pharmaceutical Scientists, the Association of Official Analytical Chemists or AOAC and the US FDA, among others).[100] The report presents guiding principles for validation of studies in both human and animal subjects that may be referred to in developing future formal guidelines. The most frequently mentioned validation criteria are given in Table 11.1.

The discussion of validation parameters in this chapter focuses on separation statistics common to chromatography and capillary electrophoresis. Many other analytical procedures, including spectroscopic measurements, dissolution testing and particle size determination, are not discussed here, although they are equally important and may be addressed in subsequent books.

The theory behind certain method validation parameters, for example, HPLC selectivity measurements using a UV-visible diode-array detector or the mathematical calculations of linearity from experimental data can be quite complex and do not fit the scope of this book. The explanations on validation parameters provided in this book are intended as guidelines only and do not give the theory and calculations used for the parameters. Where appropriate, references to other literature are made.

Selectivity

The terms *selectivity* and *specificity* are often used interchangeably. The term *specific* however, refers to a method that

Table 11.1. Parameters for method validation

- Selectivity
- Precision (repeatability, reproducibility)
- Accuracy
- Linearity
- Range
- Limit of detection
- Limit of quantitation
- Ruggedness (robustness)
- Stability

produces a response for a single analyte only[102] while the term *selective* refers to a method which provides responses for a number of chemical entities that may or may not be distinguished. If the response is distinguished from all other responses, the method is said to be selective. Since there are very few methods that respond to only one analyte, the term *selectivity* is usually more appropriate. The USP monograph[68] defines selectivity of an analytical method as its ability to measure accurately an analyte in the presence of interferences, such as synthetic precursors, excipients, enantiomers and known (or likely) degradation products that may be expected to be present in the sample matrix. Selectivity in liquid chromatography is obtained by choosing optimal columns and setting chromatographic conditions, such as mobile phase composition, column temperature and detector wavelength. The selectivity of an analytical method is tested by comparing the results of samples containing impurities to those of samples without impurities. The bias of the assay is the difference in results between the two sample types.

In biological fluids; sample matrix interference is more of a problem because it cannot be assumed that the level of interference in a blank measurement will be equal to that in a measured sample and compensated for by subtraction.

There are a variety of ways to validate selectivity. Karnes suggests that the simplest test for chromatographic analysis is to demonstrate a lack of response in the blank biological matrix.[101] As a second approach, the same author suggested checking whether the intercept of the calibration curve is significantly different from zero. Shah and coworkers[100] recommended establishing the specificity/selectivity of biological samples by using six independent sources of the same matrix.

The most difficult task is to ascertain whether the peaks within a sample chromatogram or electropherogram are pure or consist of more than one compound. In chromatography and capillary electrophoresis this can be accomplished by using spectral information, most frequently provided by mass spectrometers and UV-visible diode-array detectors. The detectors acquire spectra on-line throughout the entire chromatogram or electropherogram. The spectra acquired during the elution of a peak are normalized and overlaid for graphical presentation. If the normalized spectra are different, the peak consists of at least two compounds. The principles of diode-array detection in HPLC and their application and limitations to peak purity are described in the literature.[103–105] Examples of pure and impure HPLC peaks are shown in Figure 11.2. While the chromatographic signal indicates no impurities in either peak, the spectral evaluation identifies the peak on the left as impure.

Precision and Reproducibility

The precision of a method is the extent to which the individual test results of multiple injections of a series of standards agree. The measured standard deviation can be subdivided into two categories, *repeatability* and *reproducibility*. Repeatability is obtained when the analysis is carried out in one laboratory by one operator using one piece of equipment over a relatively short timespan.

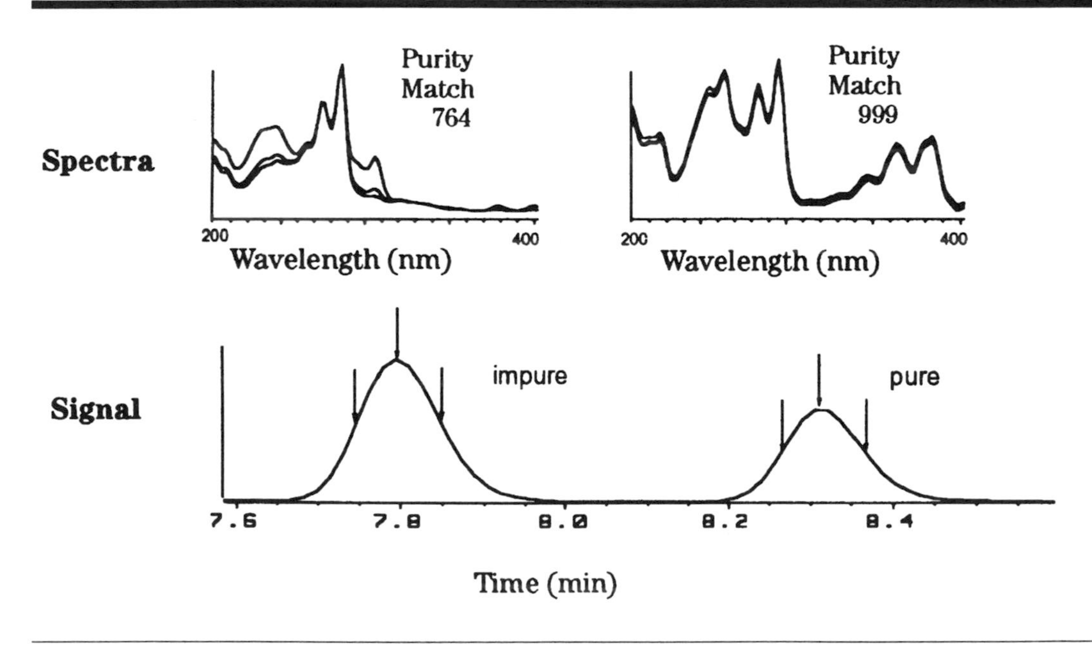

Figure 11.2. Examples of pure and impure HPLC peaks. The chromatographic signal does not indicate any impurity in either peak. Spectral evaluation, however, identifies the peak on the left as impure.[105]

Reproducibility, on the other hand, is defined as the long-term variability of the measurement process and is determined by comparing the results of a method run within a single laboratory over a number of days. A method's reproducibility may also reflect discrepancies in results obtained by different operators, from different instruments or a combination of the above. The reproducibility standard deviation is typically two- to threefold that of repeatability. Precision relative to one day is often termed *intra-day* (within one day) while that measured over a period of days is termed *inter-day* (between days).

Precision in retention time and peak area or height are major performance criteria for a separation system, especially because retention time is the primary means of peak identification and an important performance and diagnostic criterion for LC pumps and GC column ovens.

Precision of peak area is important as it is used for calculating amounts during quantification and is the most important performance criterion for an injection system. Precision should be determined using a minimum of five replicate chromatograms.[68] For bioanalytical samples precision is studied

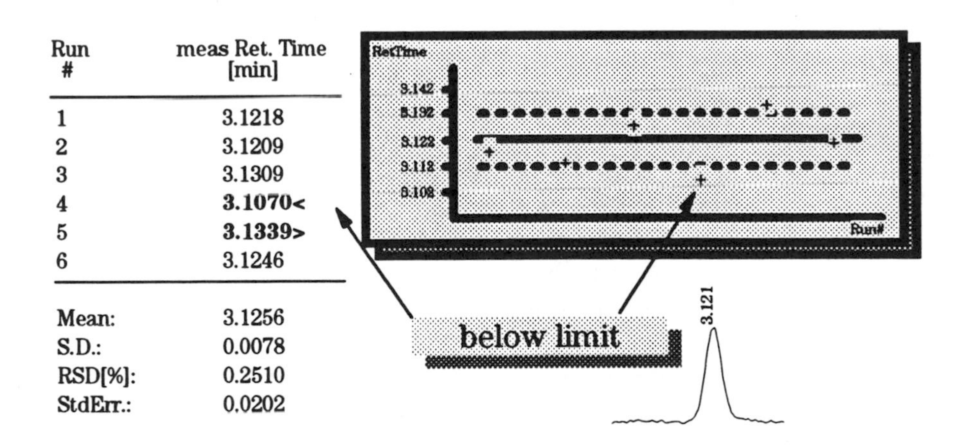

Run #	meas Ret. Time [min]
1	3.1218
2	3.1209
3	3.1309
4	**3.1070<**
5	**3.1339>**
6	3.1246
Mean:	3.1256
S.D.:	0.0078
RSD[%]:	0.2510
StdErr.:	0.0202

Figure 11.3. Commercial software is available from instrument vendors for automated method validation and system suitability testing.[92] Selected parameters and limits can be specified. Results are presented in graphical and tabular format. Exceptions to preset limits are reported.[92]

at minimum, low, medium and high concentration levels and is repeated on separate days to calculate intra-day and inter-day precision.[100]

Accuracy

The accuracy of an analytical method is the extent to which test results generated by the method and the true value agree. The true value for accuracy assessment can be obtained in two ways. One alternative is to compare results of the method with results from an established *reference method.* This approach assumes no systematic error in the reference method. A second alternative is to spike the sample matrix of interest with a known concentration of reference material. After extraction of the analyte from the matrix and injection into the analytical instrument, its response can be compared to the response of the reference material dissolved in a pure solvent.[94]

The efficiency of the extraction from the sample matrix (expressed as a percentage of the response of pure standard that was not subjected to sample pretreatment) is defined as absolute recovery.[101] Relative recovery is measured by comparing the response of the matrix (e.g., plasma) to the response of the pure solvent (e.g., water).

Linearity

The linearity of an analytical method is its ability to elicit test results that are directly, or by means of well-defined mathematical transformations, proportional to the concentration of analytes in samples within a given range. Linearity is determined by a series of injections of six or more standards whose concentrations span 50–150 percent of the assay's expected working range. The response should be linearly related to the concentrations of the standards. A linear regression equation applied to the results should have an intercept not significantly different from zero. If a significant nonzero intercept is obtained, it should be demonstrated that there is no effect on the accuracy of the method.

Furman et al.[85] recommended the following as the minimal linearity validation steps for a computerized HPLC system using the holistic validation or testing approach:

1. Calibrate the system for linearity (injector, pump, detector, data acquisition device and data output) using at least four standard solutions of different concentrations. The range should be wider than the anticipated operating range if it is known. For R&D type work where the range may not be anticipated, the full system linearity should be determined and procedures put in place for the concentration or dilution of those samples whose concentrations fall outside this range.
2. Initially run the linearity test daily.
3. If the linearity curve is reproducible over several days, the test interval may be lengthened to every other day, weekly or longer.

Range

The range of an analytical method is the interval between the upper and lower levels (including these levels) that have been demonstrated to be determined with precision accuracy and linearity using the method as written. The range is normally expressed in the same units as the test results (e.g., percentage, parts per million) obtained by the analytical method.

Limit of Detection and Quantification

The limit of detection is the point at which a measured value is larger than the uncertainty associated with it. It is the lowest concentration of analyte in a sample that can be detected but not necessarily quantified. In chromatography the detection limit is the injected amount that results in a peak with a height at least twice as high as the baseline noise.

The limit of quantification is the minimum injected amount that gives reproducible peak area measurements (equivalent to amounts), typically requiring peak heights 10 to 20 times higher than baseline noise.

Stability

Many solutes readily decompose prior to chromatographic investigations, for example, during the preparation of the sample solutions, during extraction, clean-up, phase transfer or during storage of prepared vials (in refrigerators or in an automatic sampler). Under these circumstances, method development should investigate the stability of the analytes and standards.

The term *system stability* has been defined to determine the stability of the samples being analyzed in a sample solution. It is a measure of the bias in assay results generated during a preselected time interval, for example, every hour up to 46 hours, using a single solution.[94] System stability should be determined by replicate analysis of the sample solution. System stability is considered to be appropriate when the relative standard deviation calculated on the assay results obtained at different time intervals does not exceed more than 20 percent of the corresponding value of the system precision. If on plotting the assay results as a function of time, the value is higher, the maximum duration of the usability of the sample solution can be calculated.[94]

The effect of long-term storage and freeze-thaw cycles can be investigated by analyzing a spiked sample immediately after preparation and on subsequent days of the anticipated storage period. A minimum of two cycles at two concentrations should be studied in duplicate.[100] If the integrity of the drug is affected by freezing and thawing, spiked samples should be stored in individual containers and appropriate caution should be employed for study samples.

Ruggedness

Ruggedness tests examine the effect operational and environmental conditions have on the analysis results and is the degree of variance in test results obtained by the analysis of the same samples under a variety of test conditions.

A rugged method is one that has built-in buffers against typical abuses, that is, against differences in care, technique, equipment and conditions. The ruggedness of an analytical method is determined by analyzing aliquots from homogeneous lots in different laboratories with different analysts and by using operational and environmental conditions that may differ from but are still within the specified parameters of the method (interlaboratory tests).

For the determination of a method's ruggedness within a laboratory, a number of chromatographic parameters, for example, flow rate, column temperature, detection wavelength or mobile phase cmposition are varied within a realistic range and the quantitative influence of the variables is determined. If the influence of the parameter is within a previously specified tolerance, the parameter is said to be within the method's ruggedness range.

12. Maintenance and Ongoing Performance Control

Once the installation is complete and the equipment and computer system are proven to operate correctly, the computerized system can be put into routine use. Procedures should exist which show that *It will continue to do what it purports to do.* Each laboratory should have a quality assurance program that is well understood and followed by individuals as well as by laboratory organizations to prevent, detect and correct problems. The purpose is to ensure that results have a high probability of being of acceptable quality. A plan should be set up that ensures that the system is under control. Ongoing activities may include preventive instrument maintenance, calibration, performance verification and calibration, system suitability testing, analysis of blanks and quality control samples and any combination thereof.

Maintenance

Operating Procedures for maintenance should be in place for every system component that requires periodic calibration and/or preventive maintenance. Preventive maintenance of hardware should be designed to catch problems before they occur. Critical parts should be listed and be available at the user's site. The procedure should describe what is to be done, when and the necessary qualification of the engineer performing the tasks. The system components should be labeled with the dates of the last and next scheduled maintenance. All maintenance activities should be documented in the instrument's logbook. Suppliers of equipment should provide a list of recommended maintenance activities and documented procedures on how to perform the maintenance. Some vendors also offer maintenance contracts with services for preventive maintenance at scheduled time intervals. A set of diagnostic procedures is performed and critical parts are replaced to avoid or identify problems that have not yet reached the point where they impact proper operation of the system.

Calibration and Performance Verification

Calibration

After some time, operating devices may require recalibration, for example, the wavelength of a UV-visible detector's optical unit, if they are not to adversely impact the performance of an instrument. A calibration program should be in place to recalibrate critical instrument components following documented procedures with all results recorded in the

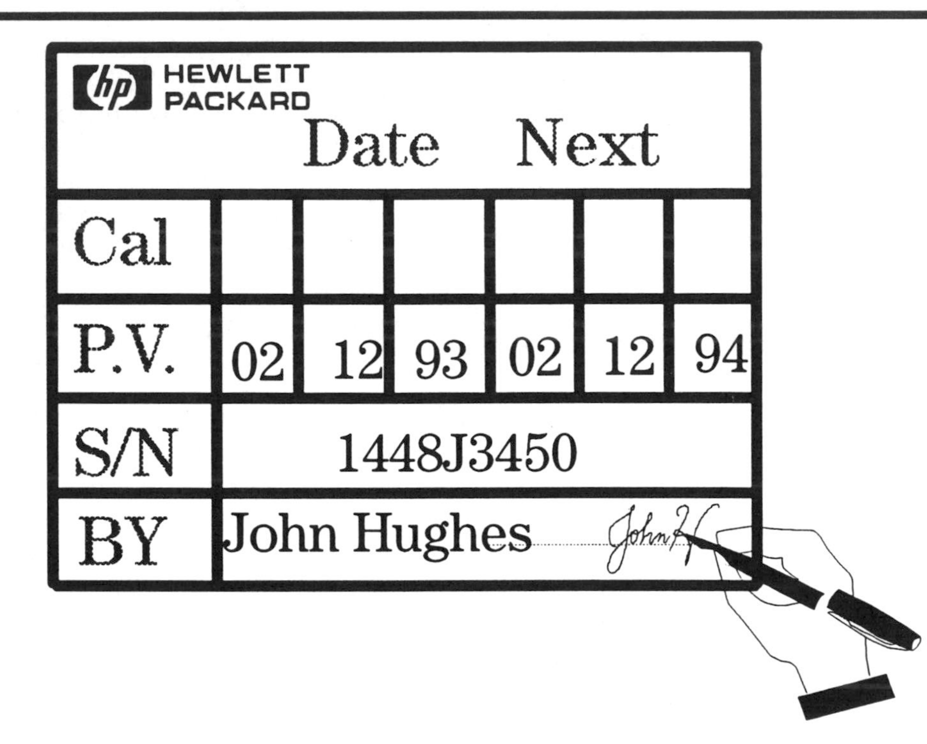

Figure 12.1. Example of a sticker with information on last and next performance verification (PV) or calibration

instrument's logbook. The computer system components should be labeled with the date of the last and next calibration. The label on the instrument should include the initials of the test engineer and the calibration report should include his or her printed name and full signature.

Performance Verification (PV)

The characteristics of equipment alter over time due to contamination and normal wear and tear. UV detector flow cells become contaminated, pump piston seals abrade and UV detector lamps lose intensity. These changes will have a direct impact on the performance of analytical hardware; therefore, the performance of analytical instruments should be verified over their entire lifetime. The *Guidance on Interpretation of the EN 45000 Series of Standards and ISO/IEC Guide 25*[60] specifies the need for performance checks in addition to maintenance and calibration: "Correct use combined with periodic servicing, cleaning, and calibration will not necessarily ensure an instrument is performing adequately. Where appropriate, periodic performance checks should be carried out (for example, to check the response, stability and linearity of sources, sensors and detectors, the separating efficiency of chromatographic systems, the resolution, alignment and wavelength accuracy of spectrometers etc.)."

The user of the equipment carries full responsibility for these activities. The supplier should provide recommendations on what to check, the procedures with test conditions, recommendations for performance limits (acceptance criteria) and recommended actions in case criteria are not met. Performance verification should follow documented procedures. Using templates is recommended for the results, an example of which is shown in Figure 12.2. Performance verification may be performed either by the user or the supplier or any other third party on behalf of the user. Whoever conducts the performance verification should have documented evidence that he or she is suitably qualified.

"Which performance characteristics should be verified and how often?" is a frequently asked question. The *Guidance on Interpretation of the EN 45000 Series of Standards and ISO/IEC*

Instrument: **HP 1050 Series VW detector**
Serial number: **1448J3450**
Test: **Baseline noise**
User's specification: **1.0×10^{-4} AU (HP's recommended limit 1.5×10^{-5} AU)**
Test frequency: **Every 12 months (HP's recom. every 12 months)**

Date	Measured value	Corrective action	Final value	Test engineer name	Test engineer signature
2/3/93	1.4x10-5			Hughes	

Figure 12.2. Example of a performance verification template

Guide 25[60] gives a recommendation on the frequency of performance checks; "The frequency of such performance checks will be determined by experience and based on need, type and previous performance of equipment. Intervals between the checks should be shorter than the time the equipment has been found to take to drift outside acceptable limits." This interpretation means that the frequency of performance checks for a particular instrument depends on *acceptable limits* specified by the user. The more stringent the limits, the sooner the instrument will drift out of the limits, increasing the frequency of the performance checks. The time interval between checks should be identified by experience and documented for each instrument.

Appendix B of the *Guidance on the Interpretation of the EN 45000 Series of Standards and ISO/IEC Guide 25*[60] lists parameters to be checked for chromatographic instruments including liquid and ion chromatographs; heating/cooling apparatus including freeze-dryers, freezers, furnaces, hot-air sterilizers, incubators, and for spectrometers, autosamplers, microscopes and electrodes. The frequency of checks for other equipment, including balances, volumetric glassware, hydrometers, barometers, timers and thermometers, is also listed.

A good recommendation is to do performance checks more frequently for new instruments. If the instrument continually meets the performance specifications the time interval can be increased.

System Suitability Testing and Analytical Quality Control (AQC)

The ongoing performance of an analytical system *for its intended use* should be controlled over time using the same analytical methods as those used for the analysis of unknown samples and using standards or reference materials with chemical structures similar to the analytes. The mechanisms proposed to prove that systems perform as expected for their intended use are system suitability tests or the analysis of quality control samples with the construction of control charts. It is recommended that users perform the checks once every day or even more often, depending on the stability of the system and the number of samples analyzed daily. The test frequency, parameters and acceptance criteria should be defined during method validation.

System Suitability Testing

System suitability tests have been proposed and defined for chromatographic systems by the United States[68] and other Pharmacopoeias. Compared to method validation, daily system suitability testing requires fewer individual determinations. A general recommendation is to check those parameters that are critical to analysis accuracy and that can change over a relatively short time. The exact type and frequency of tests should be defined during method validation. For compound analysis, the USP[68] recommends measuring:

- Precision of peak areas (system precision)
- Resolution between two compounds
- Tailing factor

Baseline noise and drift and precision of retention times are other possible parameters necessary, for example, when the detection limit is critical to the analysis.

System precision is determined by repeatedly injecting a standard solution and measuring the relative standard deviation of the resulting peak areas or peak heights. For the USP monographs,[68] unless otherwise noted, five replicate chromatograms are required when the stated relative standard deviation is 2 percent or less. For values greater than 2 percent, six replicate chromatograms should be used. For bioanalytical samples, percentage RSD should not exceed 15 percent except at the limit of detection where it should be less than 20 percent.[100]

Quality Control (QC) Samples with QC Charts

The analysis of quality control samples with construction of quality control charts has been suggested as a way to build in quality checks on results as they are being generated. Such tests can then flag those values that may be erroneous for any of the following reasons:

- reagents are contaminated
- GC carrier gas is impure
- HPLC mobile phase is contaminated
- instrument characteristics have changed over time

For an accurate quality check, quality control (QC) samples are interspersed among actual samples at intervals determined by the total number of samples and the precision and reproducibility of the method. The control sample frequency will depend mainly on the known stability of the measurement process, a stable process requiring only occasional monitoring. The WELAC/EURACHEM guide[60] states that 5 percent of sample throughput should consist of quality control samples for routine analysis and 20 percent to 50 percent for more complex procedures.

Control samples should have a high degree of similarity to the actual samples analyzed; otherwise, one cannot draw reliable conclusions on the measurement system's performance.

Control samples must be so homogeneous and stable that individual increments measured at various times will have less variability than the measurement process itself. QC samples are prepared by adding known amounts of analytes to blank specimens. They can be purchased as certified reference material (CRM) or may be prepared in-house. QC materials based on environmental matrices, food, serum or urine are commercially available for a variety of analytes. For day-to-day routine analysis it is recommended to use in house standards that are checked against certified reference material (CRM). Sufficient quantities should be prepared to allow the same samples to be used over a longer period of time. Their stability over time should be proven and their accuracy verified preferably through a comparison with certified reference material, through interlaboratory tests or by other analysis methods.

The most widely used procedure for the ongoing control of equipment through QC samples involves the construction of control charts for quality control (QC) samples. These are plots of multiple data points versus number of measurements from the same QC samples using the same processes. Measured concentrations of a single measurement or the average of multiple measurements are plotted on the vertical axis and the sequence number of the measurement on the

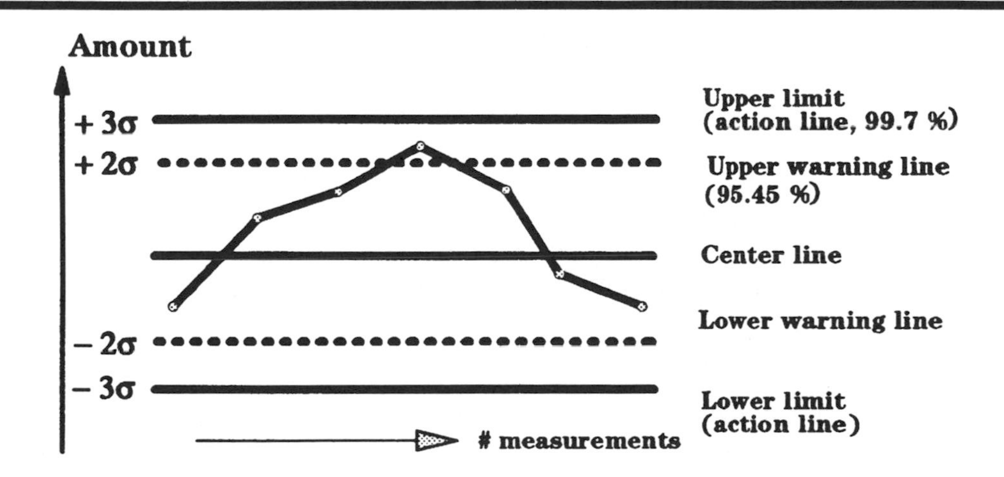

Figure 12.3. Quality control chart with warning lines and control lines

horizontal axis. Control charts provide a graphic tool to demonstrate statistical control, monitor a measurement process, diagnose measurement problems and document measurement uncertainty. Many schemes for the construction of such control charts have been presented.[106] The most commonly used control charts are X-charts and R-charts as developed by Shewart. X-charts consist of a central line representing either the known concentration or the mean of 10 to 20 earlier determinations of the analyte in control material (QC sample). The standard deviation has been determined during method validation and is used to calculate the control lines in the control chart. Control limits define the bounds of virtually all values produced by a system in statistical control.

Control charts often have a center line and two control lines with two pairs of limits: a warning line at $\mu \pm 2\sigma$ and an action line at $\mu \pm 3\sigma$. Statistics predict that 95.45 percent and 99.7 percent of the data will fall within the areas enclosed by the $\pm 2\sigma$ and $\pm 3\sigma$ limits. The center line is either the mean or the true value. In the ideal case, where unbiased methods are being used, the center line would be the true value. This would apply, for example, to precision control charts for standard solutions.

When the process is under statistical control, the day-to-day results are normally distributed about the center line and one out of 20 results may be expected to fall between the warning and action lines. No action is required if only one result falls in this area provided that the next value is inside the warning line. However, if two consecutive values fall between the warning and action lines then there is evidence of loss of a statistical control. More out-of-control situations are shown in Figure 12.4. In these cases the results should be rejected and the process investigated for its unusual behavior. Further analyses should be suspended until the problem is resolved. Instruments and sampling procedures should be checked for errors.

Quality control (QC) samples may have to be run in duplicate at three concentrations corresponding to the levels below, within and above the analysis range. For methods with linear concentration-response relationships over the full analysis range, two concentrations, one each at the high and low end of the range, are adequate.

a) one value outside the control limit
b) seven consecutive values ascending or descending
c) seven consecutive values above or below the centerline
d) two out of three consecutive values outside the warning limits

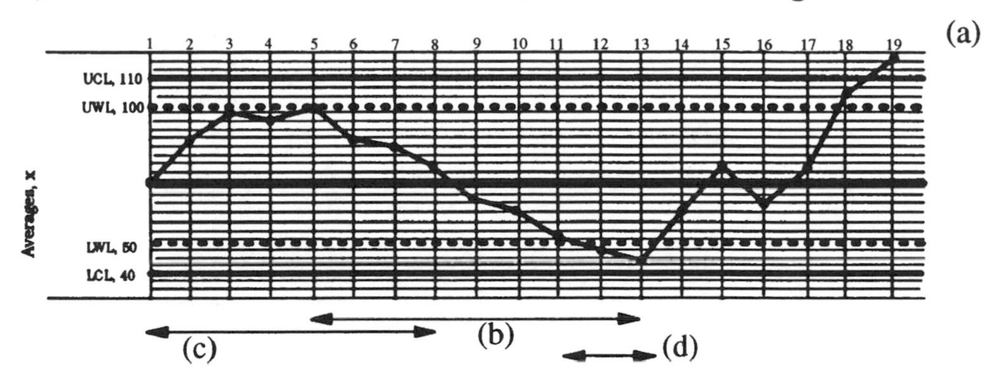

Figure 12.4. Possible out-of-control events

A documented quality procedure should be in place that provides the operator with step-by-step instructions in the event that the results of one or more QC samples are outside the warning or control line. There are two types of corrective action, immediate on-the-spot and long-term. On-the-spot action is used to correct minor problems, such as the replacement of defective instrument parts, an HPLC UV-visible detector lamp for example. These actions can be performed by a single individual, and analytical methods or procedures do not need to be changed.

Long-term action is required when an out-of-control situation is caused by a method, an uncommon equipment failure or laboratory environment problem. For long-term actions, one person is made responsible to investigate the cause, develop and implement corrective action and verify that the problem has been solved.

Handling of Defective Instruments

Recommendations on the handling of defective instruments can be found in the ISO/IEC Guide 25.[59] Clear instructions should be available to the operator on actions to take in the event an instrument breaks down or fails to function properly. Recommendations should be given on when operators should attempt to rectify the problem themselves and when they should call the instrument vendor's service department. For each instrument there should be a list of common and uncommon failures and every problem should be so classified. Common problems like a defective UV-visible detector lamp require a short-term action. The lamp should be replaced, and after a functional test, the instrument can be used for further analyses. The failure, repair and result of the functional test should be entered into the instrument's logbook.

In the case of an uncommon failure that cannot be easily classified and repaired by the operator, several steps are recommended.

- The problem should be reported to the laboratory supervisor or to the person responsible for the instrument, who will decide on further action.
- The instrument should be removed from the laboratory and stored in a specified area, or if this is impractical due

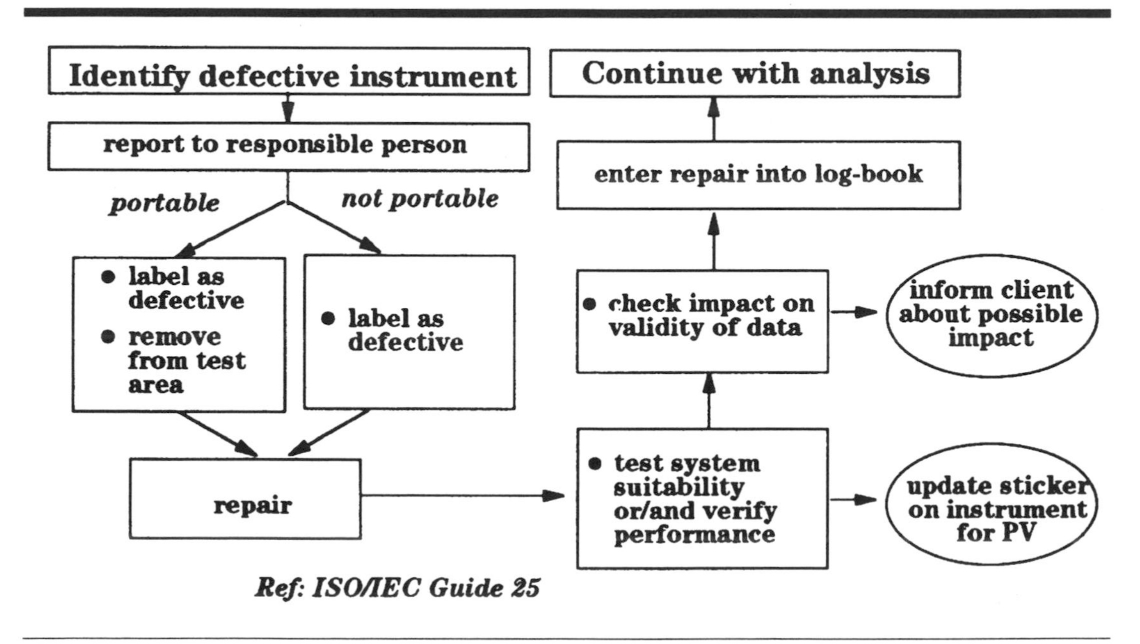

Figure 12.5. Handling of instruments with uncommon failures

to its size, it should be clearly labeled as being defective. For example, portable equipment like pH-meters should be removed while larger equipment like an HPLC, a GC or an ICP-MS system should be labeled "out of service".

- After repair, correct function must be verified, the type and extent of testing depending on the failure and possible impact on the system. Depending on the failure, this may require part of full performance verification (requalification) or only system suitability testing.
- Impact of the defect on previous test results should be examined.
- Clients should be informed about the effect the failure could have had on the validity of their data.
- An entry on the defect, repair and performance verification should be made in the instrument's logbook.

13. Testing of Chromatographic Computer Systems

Chromatographic computer systems perform different tasks including instrument control, data acquisition, peak integration, peak identification, calculation of quantitative results and reporting, storage and transfer of data to other programs on the same computer or via networks to other computers. Not all chromatographic computer systems perform all the tasks listed above. While most perform data acquisition, peak integration and quantification, some also control spectrometric detectors, evaluate data and provide capabilities such as peak purity checking and identification.

Chromatographic computer systems are tested extensively at the vendor's site prior to release to manufacturing. Before a computer-controlled analytical system is put into operation at the user's site, acceptance testing should be carried out to make sure that the system does what it purports to do in a typical user's environment.

The OECD GLP principles[50] state that validation can be done by the supplier on behalf of the user at the time of software development, but the user should undertake formal acceptance testing: "It is the responsibility of the user to ensure that the software program has been validated" and "It is acceptable for formal validation of applications software to be carried out by the supplier on behalf of the user, provided that the user undertakes the formal acceptance tests."[50]

In its Good Automated Laboratory Practices, the US EPA recommends specifying acceptance criteria for any software and the development of SOPs for acceptance testing; "Acceptance criteria should be specified for new or changed software. Written procedures should be in place for acceptance testing on what should be tested, when and by whom. Test results should be documented, reviewed and approved."[20]

Acceptance tests should be performed by the user in a typical user environment regardless of whether the software was developed by the user's firm or by an instrument manufacturer. The acceptance criteria should be specified for new or changed software before testing begins. Documentation of such testing should include the acceptance criteria, summary of results, names of persons who performed the testing, an indication of review and written approval. Only if the acceptance criteria are met and approved, should the system be placed in service. A frequently asked question is what and how much acceptance testing should be done. The type and number of tests should be based on a good scientific judgment and should be selected such that the system has proven to provide accurate, precise and reliable results.

Most analytical computer systems provide far more functionality than is needed for a specific application. Acceptance testing should be limited to those functions of the system that are used in a laboratory. For example, software purchased to acquire and evaluate multiple signals from a UV-visible diode-array detector usually includes the capability for spectral evaluation. If this part of the software is not going to be used, it does not need to be tested. On the other hand, for some applications the user may add software in the form of MACROs, for example, for further customization of data evaluation and reporting. This is considered to be software that not only has to be acceptance tested but also has to go through a formal software validation process.

Instrument vendors should provide a functional specifications list for each computer system. Before writing the protocol for acceptance testing, the user should define which of the listed specifications will be used in routine analysis, and only the items on this list should be tested. Vendors should also provide test routines that perform most important tests automatically. For example, for a chromatography data system these functions should include verification of communication to the analytical equipment, data acquisition, peak integration, quantitation, reporting as well as file storage and retrieval.

In this chapter special considerations for testing chromatographic computer systems are discussed. Examples of specifications and a template for manual modular functional software testing, the concept of integrated software testing and a procedure of automated verification of signal integration and reporting are presented. Modular testing is more time consuming and should always be done at the vendor's site. It is recommended at the user's site for diagnostic purposes or if there is little or no experience with the analysis technique. Integrated system testing is done at the vendor's site and is recommended for users who already have experience with the instrumentation and test results from existing instruments for a comparison.

Examples for Specifications and Tests for a Computerized HPLC System

Specifications

Before any testing to verify the performance of a chromatographic computer system is undertaken, the system's functional specifications should be defined. An example of *selected* specifications for an HPLC system taken from the literature[107] are given in Table 13.2.

Modular Functional Testing

Next, the functions that should be performance tested are defined. For testing the individual items, development of templates that include a cover template with information on test environment, instrument configuration and software revision is recommended. It is important that the templates include

Table 13.1. Typical tasks of chromatographic software with testing requirements

☑	Instrument control	☑	Quantification of results
☑	Data acquisition	☑	Report generation
☑	Peak identification	☑	Data storage and retrieval
☑	Peak integration	☑	Data transfer through networks

Table 13.2. Example for specifications of an automated HPLC system

Instrument control

Example—fluorescence detector

The excitation and emission wavelength of the HP 1046A fluorescence detector can be set from 190 to 800 nm, in steps of 1 nm. Gain, response time, gate and lamp frequency may also be set. Gain and changes in the emission and excitation wavelength may be time programmed. The excitation and emission wavelengths may be optimized by analyzing scans.

Data acquisition

Example—instrument status display

The status of the instrument is continually monitored and updated on the display, along with the elapsed run time of the analysis

Data analysis

Example—manual or "rubber band" integration

Manual or "rubber band" integration is available for users whose chromatography requires human interpretation. This functionality enables users to graphically define peak start and stop points and have the recalculated areas incorporated in the quantification and reporting peaks. Such manually reintegrated peaks are identified on reports..

Data analysis

Example—quantification

Quantification is based on the *area percentage, normalized percentage, external standard* and *internal standard* calculations. Calibrations may be multilevel and multiple internal standard types.

Data analysis

Example—report device and output format

The report device and output format for a file may be specified. When a file is specified as the destination device the software supports ASCII, comma separated values (CSV), data interchange format (DIF) and Windows metafile (WMF) output formats.

Automation

Example—multimethod sequencing

The ChemStation can execute multimethod sequences for 100 different samples from the HP 1090 autosampler or 119 samples from the HP 1050 Series autosampler.

Continued on next page

Continued from previous page

Others

Example—recording of actual instrument conditions for GLP compliance

The actual instrument conditions (pressure, flow, gradient and temperature) that occurred during each analysis are recorded. These data can be subsequently displayed graphically with the chromatogram to show the actual instrument conditions during that particular analysis, and can be included in the report.

entry fields for observations of actual results made during testing, as a simple pass/fail indication is insufficient. Tests that check the system's error handling capability should be included. Wrong entries, for example, a flow rate entry above the operational range, should be recognized as an error and displayed. Another simple test is to see how the program responds when alpha data is inputted to an entry field that expects numeric data. The test should also check the boundary conditions of the software. To test these conditions, data is inputted that is slightly above or below the operational limits. For example, if the operational limit of a GC oven is 400°C, entries of 399°C and 401°C should be tested.

Figure 13.1 shows a simple template with an example for manual testing. The purpose of the test is to verify proper communication between the computer and the UV-visible detector of an HPLC system. The test should also demonstrate that the systems respond correctly to data entries close to and above the upper wavelength limit. This type of test is also recommended for systems without built-in error handling or if the behavior of the error handling is suspect. The templates include areas for date of the test, test item, test description, expected results, actual results (observations), pass/fail checkboxes and entry fields for comments and signatures.

Integrated System Testing

Alternatively to modular functional software testing the integrated approach can be applied. The system software is tested with a few experiments using test samples during a complete analysis run. This type of testing is recommended if the analysis technique is well known in a laboratory by the operators. The new system should perform the same functions as

Figure 13.1. An example template for modular functional software testing

Date	May 27, 1993
Test item	Instrument control: Programmability of the detector wavelength
Test description	1. Set the wavelength on the HP 1050 Variable Wavelength Detector to 280 nm. 2. Set a time program. at 1.0 min. 225 nm (normal operating range) at 1.2 min. 599 nm (close to limit of 600 nm) at 1.4 min. 680 nm (above limit of 600 nm) 3. Press "Start Run" [ENTER].
Expected result/ Acceptance criteria	No error message. After 1.00 min. the wavelength should switch to 225 nm. Result should be displayed. No error message. After 1.2 min. the wavelength should switch to 599 nm. Result should be displayed. Error message on computer screen after setpoint entry: out of range After 1.4 min the display should continue to show 599 nm.
Actual result (observations)	*1.00 minute after the start the wavelength switched to 225 nm. This was shown on the display.* ☑ pass ☐ fail *1.20 minutes after the start the wavelength switched to 599 nm. This was shown on the display.* ☑ pass ☐ fail *An error message appears on the screen: out of range.* *1.40 minutes after the start the display continuous to show 599 nm.* ☑ pass ☐ fail
Comments	*Test passed*
Test engineer	Name: M. Burger Signature: *M. Burger*
Reviewed by	Name: K. Romero Signature: *K. Romero*

existing ones and well characterized reference standards or control samples and test chromatograms should be available which can be used to compare results obtained from the new equipment with that of existing ones. It is recommended to develop a test method where all key functions are executed, for example, instrument control, data acquisition and peak integration. Deviations from expected results are recognized and modular testing can be used to diagnose the source of the problem. The reference standard or control sample should be well characterized and the expected chromatograms or spectra familiar to operators.

A widely used practice is to test the software and computer system together with the analytical equipment. Equipment tests usually include functional and performance testing and many key software functions such as instrument control, automation, data acquisition, integration, calibration, reporting and data storage and retrieval are executed. When all the tests are successfully completed and criteria for equipment functions and performance tests are met also the software and computer systems can be assumed to be 'fit for its use'.

Verification of Peak Integration

The most difficult task for chromatographic software is to perform correct peak integration on asymmetric, poorly resolved or unresolved peaks, tailing peaks, peaks on a drifting baseline and peaks with a signal-to-noise ratio close to the detection limit.

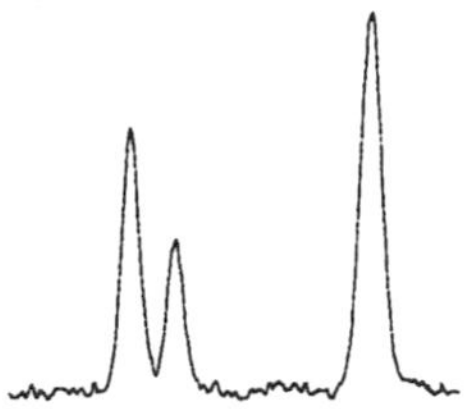

Chromatography is a relative method. Sample amounts are calculated by comparing unknowns with the response of a standard. To test an integrator for correct peak integration and quantification in operational situations, one or more standard test chromatograms with known dimensions should be available. The integrator measures the standard, and the results can be compared with what they are known to be. A well characterized test chromatogram, an integration method and calibration table and a printout of the expected results should be supplied together with the software.

Test chromatograms used routinely should reflect typical sample chromatograms. They should include well-separated peaks if all sample compounds are resolved and unresolved peaks when sample compounds are not completely separated. Drift is also important; if samples display significant baseline drift, the test chromatogram should also include drift.

The user should be able to define and use his or her own chromatogram for the test. One possible solution would be to connect the data system to a measurement device, a GC or HPLC system, for example, inject a test sample that closely represents a typical unknown sample and compare the results with previously recorded chromatograms acquired from the same instrument. The problem here is that the results will not depend entirely on the integrator but also on other variables, such as stability of the test sample, precision of injection volume and flow rate, column performance and detector response. In practice, it is impossible to generate exactly the same digital numbers if the chromatogram is generated from a real sample. Done in this way, the precision of the analytical instrument will be less than that of the chromatographic data acquisition and evaluation system. Verifying the integrator's integration of shoulders, for example, is difficult on an HPLC system, because the shoulder may not be reproducible over long periods of time.

A second method utilizes electronic peak generators to produce reproducible peaks. They generate reproducible outputs, usually gaussian peaks, which can be played and replayed as often as required. The problem with this method is that the output is fixed and will not usually represent chromatographic peaks from real-life samples with shoulders, tailing and so on. Caution should also be exercised when using synthetic signals for validation as they do not have the same noise levels as real samples. Furthermore, an additional hardware device that must be independently validated is needed.

Dyson[108,109] recommended using synthetic chromatograms generated by computer software to verify the performance of an integrator. The signals are fed from the computer, via a special D/A card, to the integrator's analog input and thereby into the integrator. The advantage of this procedure is the permanent availability of repeatable and reproducible chromatograms with accurately known dimensions (area, height, retention time, asymmetry, tailing). The results generated by the integrator can be compared with the previously characterized, absolutely accurate results. A single computer can be used to verify the performance of all integrators in a laboratory, and the same software may be used to verify all types of integrators within a company or for interlaboratory tests. With a graphics package, peaks with different shapes may be synthesized representing different real-life sample peaks with peak tailing, drifting and noisy baselines.

An alternative method is proposed for Hewlett-Packard's ChemStations. Correct peak integration is verified as part of an automated ChemStation performance verification program using either well characterized standard chromatograms or chromatograms derived from application specific real-life samples. The system will match newly calculated integration and quantification results with ones that have been previously generated and archived in unalterable, checksum-protected binary register file. The procedure is described in the next section.

Automated Procedure for Testing a Chromatographic Computer System

The procedure can be used to verify the performance of Hewlett-Packard's ChemStations for formal acceptance testing, operational qualification or requalification

- at installation,
- after any change to the system (computer hardware, software updates,
- after hardware repair or
- after extended use.

Successful execution of the procedure ensures that

- program files are loaded correctly on the hard disk,

- the actual computer hardware is compatible with the software and
- the actual version of the operating system and user interface software is compatible with the ChemStation software.

The method tests key functions of the system, such as peak integration, quantitation, printout, and file storage and retrieval.

Test chromatograms derived from standards or real samples are stored on disk as master file. Chromatograms are supplied as part of the software package or can be recorded by the user. This *master data file* goes through normal data evaluation from integration to report generation. Results are stored on the hard disk. The same results should always be obtained when using the same data file and method for testing purposes. The software includes a routine to carry out the performance tests and verification automatically.

To generate the master file, the user selects and defines a *master chromatogram* from a file menu and defines a method for integration and calibration. The instrument will perform the integration of the peaks using the specified method and stores the results in a checksum-protected, binary data file for use as a reference in later verifications. For performance verification the user again selects the same master chromatogram and the same method. The program integrates and evaluates peak data, compares it to the master set stored on disk, then prints a report informing the user of the successful verification of the ChemStation.

The advantages of this method are several fold:

1. A user can select one or more data files that are representative of the laboratory's samples. For example, one file may be a standard with nicely separated peaks that cover a wide range of calibration concentrations, useful to verify the accuracy and the linearity of the integrator. Second or third chromatograms can be used to verify the integrator's capability to integrate real-life samples reproducibly. Chromatograms may be selected with
    - peaks close to the detection limits,
    - peaks with shoulders,

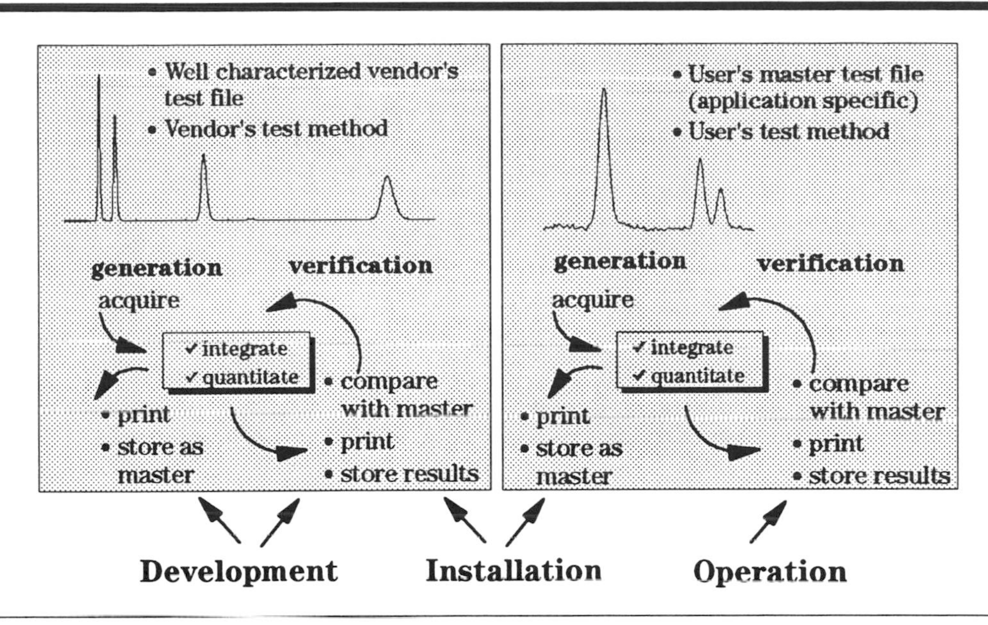

Figure 13.2. Verification process of chromatographic evaluation software. A standard chromatogram is generated by the vendor and is well characterized and documented while the software is being developed. The results are verified at installation to make sure that the system works as intended at the user's site. The user can then generate his or her own application-specific *master* chromatograms to verify the ongoing performance of the system.

- poorly resolved peaks,
- peaks on drifting baselines,
- peaks with tailing or
- asymmetric peaks.

2. The entire process is fully automated avoiding erroneous results and encouraging users to perform the test more frequently.
3. Results of the verification are documented in such a way that the documentation can be used directly for internal reviews and external audits.
4. Detailed test data can be printed on request.

Figure 13.3. An HP ChemStation verification report

=====================================

ChemStation Verification Report

=====================================

Tested Configuration

Component	Revision	Serial Number
Diode-array detector	1.0	3148G00859
HPLC 3D ChemStation	Rev. A.02.00	N/A
Microsoft Windows	3.10 (enhanced mode)	N/A
MS-DOS	5.0	N/A
Processor	i486	N/A
CoProcessor	yes	N/A

ChemStation Verification Test Details

Test Name : C:\HPCHEM\1\VERIFY\CHECK01.VAL
Data File : C:\HPCHEM\1\VERIFY\CHECH01.VAL\DEMODAD.D
Method : C:\HPCHEM\1\VERIFY\CHECK01.VAL\DEMODAD.M
Original Acquisition Method : PNASTD
Original Operator : Hans Obbens
Original Injection Date : 08/02/1985
Original Sample Name : AIRTEST

Signals Tested
Signal 1 : DAD B, SIG=305, 190 Ref=550,100 of DEMODAD.D
Signal 1 : DAD B, SIG=270, 4 Ref=550,100 of DEMODAD.D
Signal 1 : DAD B, SIG=310, 4 Ref=550,100 of DEMODAD.D

ChemStation Verification Test Results

Test Module	Selected For Test	Test Result
Digital electronics test	Yes	Pass
Integration test	Yes	Pass
Quantification test	Yes	Pass
Print analytical report	Yes	Pass

ChemStation Verification Test Overall Results: Pass

HP 1050 LC System, Friday, June 18, 1993 12:16:55 PM by ________________

Page 1 of 1

14. Data Validation, Audit Trail, Security and Traceability

Data validation is the process by which data are filtered and accepted or rejected based on defined procedures[110] and is the final step before release. Standard operating procedures should exist for the definition of raw data, data entry security and review. The plausibility of critical data should be checked, no matter if data were entered manually or were transferred electronically from an analytical instrument. Checks, preferably performed automatically, should be built into any routine method to identify errors. Requirements for a validity check of data include well-maintained instruments, documented measurement processes and statistically supported limits of uncertainty.

Three major steps required for data verification and validation are illustrated in Table 14.1.

Final results should be traceable back to the individual who entered the data, or where data is acquired on-line from an analytical instrument, the instrument should be identified. In the latter case it is recommended to store the instrument serial number, method parameters and the instrument conditions together with the raw data. Any failure or unforeseen event that occurred with the instrument should be recorded automatically in a logbook and stored together with the raw data. The impact of the error on the data should be evaluated and suitable action taken. If changes to any data have been made, the original raw data must not be obscured. The person who made the change must be identified, and the reason for the change should be given, together with the date.

Table 14.1. Phases of data validation

Data entry

- ☑ Identify person for manually entered data
- ☑ Identify instrument for electronically entered data
- ☑ Verify accuracy of critical data (formally required by some regulations)

Data change

- ☑ Audit trail
  Specify name of authorized person who changed the data, reason for change and date of change

Validation of processed data

- ☑ Check plausibility
- ☑ Accept or reject data

Data Entry

Data entry plays a major role in the security of computer-processed data and, although stringently regulated, is probably the area most often violated. "Data integrity is most vulnerable during data entry whether done via manual input or by electronic transfer from automated instruments."[20] To comply with all the guidelines requires a major appraisal of the role of computer system operators and will often require new software functionality.

Official directives and recommendations from the European Union and the US EPA include paragraphs on data entry. For example, Annex 11 of the EU GMP directives includes; "When critical data are entered manually there should be an independent check on the accuracy of the record which is made. The system should also record the identity of the operator(s) involved."[39] The US EPA GALP recommendations also have a paragraph on data entry; "The laboratory shall have written procedures and practices to verify the accuracy of manually entered and electronically transferred data collected on automated system(s)."[20]

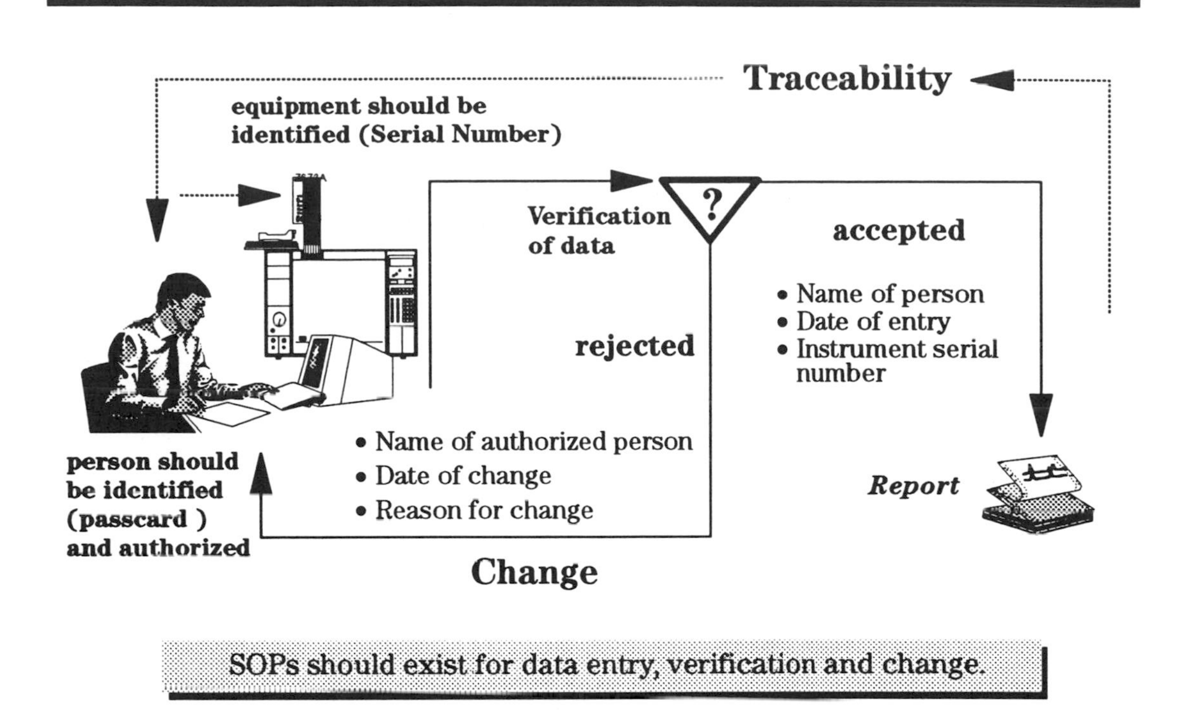

Figure 14.1. Data entry, verification and change

Manual Data Entry

There is a difference between entering data manually into a laboratory notebook and entering data into a computer. With a notebook, reading and writing errors usually occur when the writing style of the person making the entry is poor. This can usually be corrected by coaching the person. Computer typing errors on the other hand may occur at any time, can be very difficult to detect and verify and can have a severe impact on the data. An incorrectly entered amount in a sample table will give an erroneous analysis result which is difficult to detect.

Which data should be verified in a chromatographic control and data system? The recommendation is to verify only those data that are critical to the final result and that can never be verified later. For example, if a sample weight is transcribed directly from the display of a balance into the computer's sample table, the data entry should be verified. On the other hand, the setpoint entered for a gas chromatographic column oven temperature becomes part of the method that can be printed out later and compared with the original standard operating procedure, obviating the need for data entry verification. It is recommended to have builtin security mechanisms for critical data entries. Currently there is no good solution for the verification of manual data entry. Double keyboard entry is very unpopular and frustrating and still does not always reduce the expected errors to an acceptable level. Reducing human input, e.g., through connecting a balance directly to a computer, is a preferred solution, but not always possible.

Tracking Equipment and People

Final analysis results should be traceable back to the individual who entered the data. "The individual person responsible for direct data input shall be identified at the time of data input."[20] and "Data should only be entered or amended by persons authorized to do so."[39] The individual responsible for direct data input should be identified at the time of input. This can be accomplished by forcing the operator to enter her or his name or a personal code (password) into the computer system before any data entry can be made.

If data are entered directly from analytical instruments, the instrument transmitting the data shall be identified, together with the time and date of transmittal. This can be accomplished by entering the instrument's serial number along with a date and time stamp into each data set transmitted to the computer.

Raw Data: Definition, Processing and Archiving

Traceability of reported data back to raw data is important for manual and automated computer systems. For GLP studies raw data must be archived.

Definition of Raw Data

The definition of raw data in manual operations is uncomplicated compared to computerized data and has been defined by the US FDA GLP regulations (58.3 part k); "Raw Data means any laboratory worksheets, records, memoranda, notes, or exact copies thereof, that are the result of original observations and are necessary for the reconstruction and evaluation of the report of the study." Raw data must be collected promptly, directly and legibly and must be signed and dated.[111] They should be identified and collected as specified in the protocol and standard operating procedures. They must be recorded accurately, verified and eventually compared to the information in the final report. Raw data tend to be the first notes from an observation that are written down in a laboratory notebook or on a form. Examples include results of environmental monitoring, instrument calibration records, maintenance records, sampling conditions and integrator outputs from analytical equipment. Raw data may also be entries in the worksheet used to read and note information from the LED display of an analytical instrument.

If manually recorded data are transferred to a computer database, neither the electronically stored data nor its paper printout can substitute for the original.[111] If laboratory observations are entered directly into the computer, the storage media is considered to be the raw data. For data captured directly by a computer, for example, when a balance is connected to a computer, the laboratory may elect to treat either the electronically recorded information or a hard copy printout as raw data. The correct definition of raw data should be described in an SOP. If a hard copy is retained, the magnetic copy may be deleted. Magnetic media treated as raw data must be capable of being displayed in readable form for the entire period that the information is required to be retained. The data should also be protected against unauthorized change.

Defining and Archiving Chromatographic Raw Data

Chromatographic raw data may be defined as either the first paper printout or as the electronic data originally captured by the data system. In the case of a simple integrator, the raw data format will be the first paper printout, because that is all that is available. Raw data on paper should usually include chromatograms and printed analysis reports with peak areas and/or peak heights, retention times and the amounts for

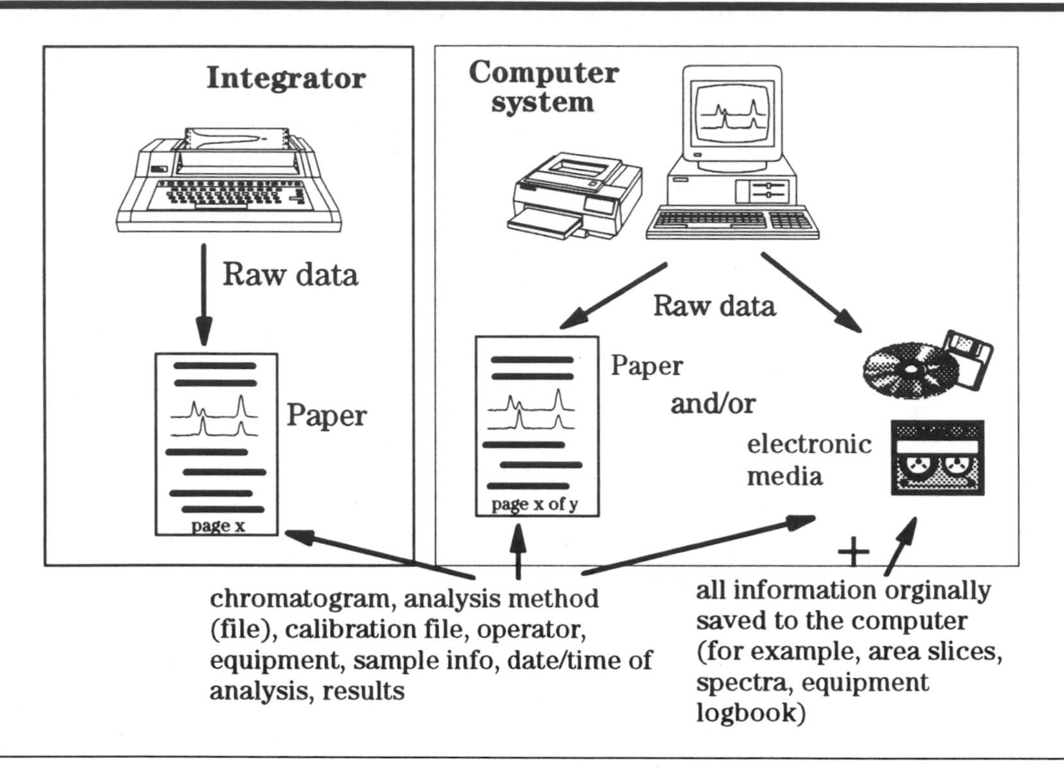

Figure 14.2. Possible definitions of raw data in chromatography

quantitative results. The paper documentation should also provide information on who performed the analysis, when, where, how it was done (reference to method files) and on which equipment. This information may be inscribed manually or added automatically by integrators having this capability.

Where a sophisticated computer system is used, either the first computer printout or the electronically archived digital information may be defined as raw data. If paper printouts are used as raw data, they should include chromatograms with integration marks, result reports, sample information and reference to the analysis method. Modern chromatographic workstations employ special report generators that allow all types of information to be printed and plotted automatically at the end of each analysis run. An example of a paper printout template with relevant raw data information is shown in Figure 14.3.

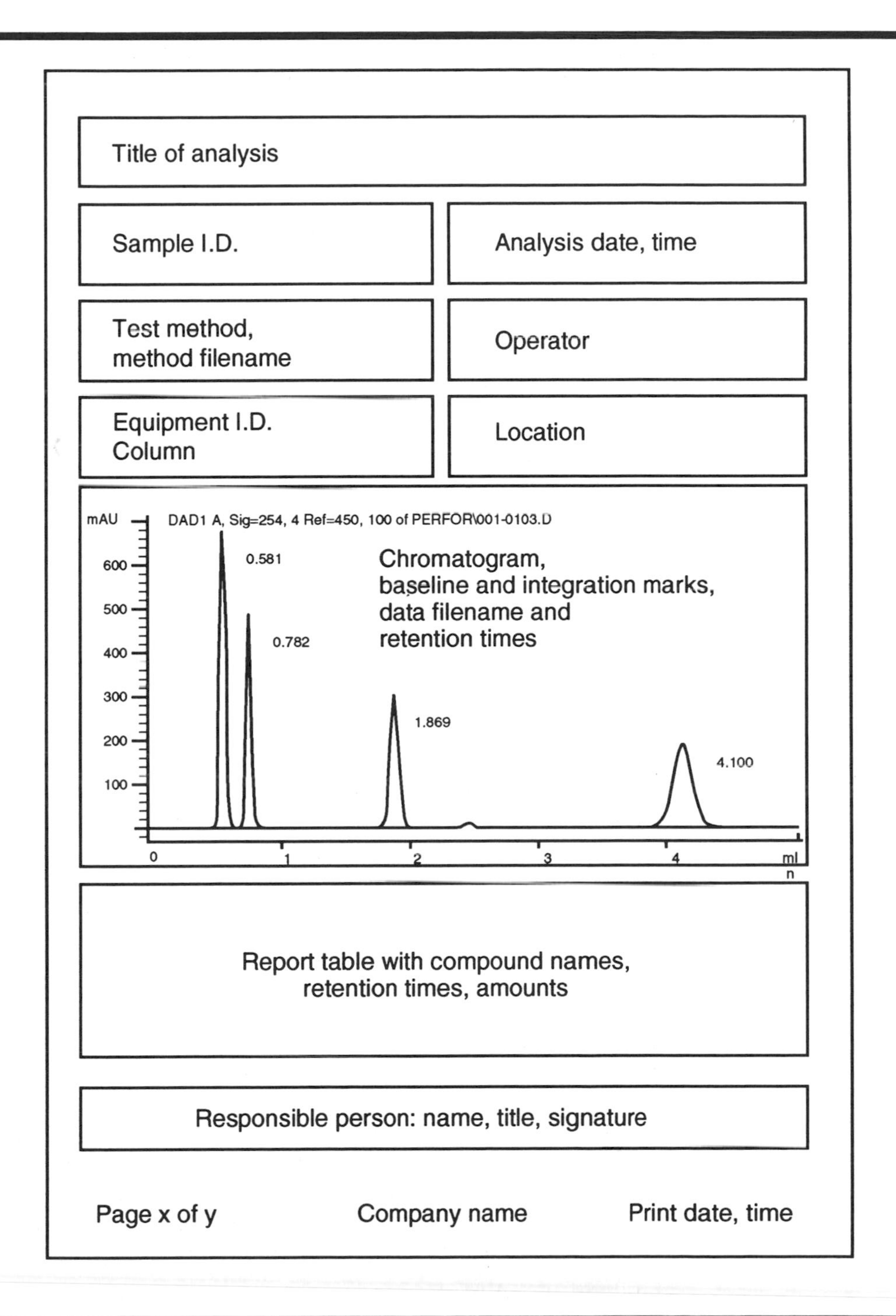

Figure 14.3. Chromatographic raw data can be defined as the computer printout.

The same information can be saved electronically as a report file and the storage medium be defined as raw data. The advantage would be that more information could be saved in less space.

It is also possible to define digital data originally captured by the data system as raw data. In chromatography and capillary electrophoresis these are area slices between the electronic zero line and the chromatogram or electropherogram at different runtime intervals of typically 5 to 50 Hz. Even though these numbers usually can be printed they can hardly be used for direct evaluation by most operators as long as they are not graphically displayed and converted to more meaningful numbers such as peak areas or amounts. Besides that it is difficult and frequently impossible to reconstruct these data mathematically (area slices) from processed data (peak areas or amounts) because integration algorithms are propriety of the vendor and typically not known to the user. On the other hand, if analytical method, integration, calibration and report are saved together with the raw data, the original chromatogram or electropherogram can be reconstructed easily.

Where computer workstations are used, Furman and his colleagues[85] recommend storing as much data as possible on electronic media. For example, where a diode-array detector is used for an HPLC analysis they recommend storing all spectral data, enabling additional information to be obtained later if required. For instance, if a peak appearing in the chromatogram was later found to be of significant importance in a GLP study, spectral data stored on disk or tape could be used to identify the peak as the compound of interest in all previous chromatograms avoiding the necessity to rerun all the samples, assuming they were still available. The authors also provide reasons for storing the data on paper:

1. There is some doubt as to whether two different operators could generate exactly the same results from a set of raw data. There are too many choices on a computer system that influence this process, for example, to smooth data, to reject noise levels or to define start and stop points for peak integration.

2. If the computer system becomes obsolete, it may be impossible to regenerate final data from raw data.

3. They believe that internal and external reviews by quality assurance staff and FDA investigators would be facilitated by the availability of the paper records that documented final decisions.

If data have to be archived for relatively short periods, the first concern above can be solved by storing all integration and calibration tables as well as report parameters as method files together with raw data. This requires that the same software, preferably with the same revision number, also be archived. This can be quite complicated should the data need reprocessing after several ears. Not only should the exact software revision be reloaded, but also the original computer operating system is required. It may also be difficult to find operators sufficiently familiar with the software to reprocess the data. Finally, Furman's[85] second concern about the availability of hardware can become a critical issue when data must be archived for more than 10 years.

The need to have old hardware and software is avoided when vendors provide validated programs to convert original raw data into a format that can be processed by current computer systems and software.

Because of all these problems, it may be easier to define and archive chromatographic raw data on paper. On the other hand, electronic data include valuable information that may be useful at a later date. A good compromise may be to define paper as raw data for archival over the entire retention period as required by clients or regulatory agencies and in addition to store electronic data, but only for as long as they could be of scientific interest for the study or other ongoing projects.

If electronic data are defined as raw data, the following points should be considered:

1. Data should be safely stored over the entire retention period. It is certainly not enough to store data on a single tape and expect that they will still be readable in 20 years time. The recommendation is to store the data twice, preferably on optical disk drives and to reload the files at certain time intervals, for example, every five years, to check the file integrity and store the files on new media.

2. Data must be available for reprocessing and/or printing over the entire retention period. The problem here is that software vendors will rarely give any assurance that compatible programs will be available for all future revisions of their software or replacement products, and such assurances would anyway be of little help should the company go out of business. A good recommendation is to look at the history of the company and ask the question, "Which vintage of data files can be processed with today's software?"

3. Final data can be processed from raw data in exactly the same way it was processed during the first analysis. This can be accomplished by saving all chromatographic, integration and calibration parameters, and if a special report formatting program was used, it should also be saved.

4. The integrity of the data must be ensured. This can be accomplished by using software that stores data in a binary file format, making intentional manipulations extremely difficult and unpredictable. Checksum procedures, which check file integrity, should be part of the data archival and retrieval routines ensuring that any accidental damage to data files is reported.

5. Care should be taken to ensure that there are no electromagnetic fields in the vicinity of the storage media when data are archived electronically.

Audit Trail for Amended Data

When changes are made to raw data, for example, when an area is converted to an amount, the original area data must not be overwritten or deleted. Similarly, if chromatographic area slices are treated as raw data, this file must not be deleted after the integrated results have been calculated, printed and archived. A specific paragraph on amended data is included in the US FDA GLP regulations; "Any change in automated data entries shall not obscure the original entry, shall indicate the reason for the change, shall be dated, and shall identify the individual making the change" (US FDA GLP 58.130 part e).

In chromatography and capillary electrophoresis this means that when raw data files are integrated, the integration results and other processed data must not overwrite the raw data file. Intermediately processed but rejected data need not be archived. This is important during a research analysis, for example, when a few integrations with different conditions may be necessary to find the correct integration parameters. In this case only the last and final integration parameters, the final results and the original raw data should be archived. Preliminary integration parameters and results can be deleted.

If during a routine analysis an established and documented integration method is changed, the reason for the change should be documented, the person who made the change should be identified and the new parameters should be archived together with the new results.

In this case the following information should be archived:

- raw data
- original integration parameters
- original processed data
- final integration parameters
- finally calculated integration results
- the date and time of the change
- who made the change
- the reason for the change

If electronic formats are used for archival, it is recommended that the instrument parameters, the instrument conditions and the instrument's logbook together with the raw data be electronically archived. This will facilitate the exact reconstruction of the original analysis.

Validation of Data

Critical data should be validated by a qualified and authorized person following a standard operating procedure. A prerequisite for accurate analysis data is a correctly functioning instrument. Preventive maintenance together with regular calibration and performance verification will facilitate the instrument's ability to generate accurate data. Checks should be made for proper sample identification, transmittal errors and consistency.[110] Techniques to accomplish this include intercomparisons with similar data, checks for plausibility of values with respect to specified limits, regression analysis and tests for outliers. Checks, preferably automated, should be built-in to any routine method to identify errors.

In chromatography and capillary electrophoresis, peak shape, resolution and integration marks should be checked to ensure that peaks are suitable for quantitative analysis.

Frequently discussed questions are how and how many data should be validated? The answer depends on the analysis task, the analysis method and on the probability of getting wrong data. For example, if a soil sample is analyzed for organic compounds with HPLC and UV detection, and if the

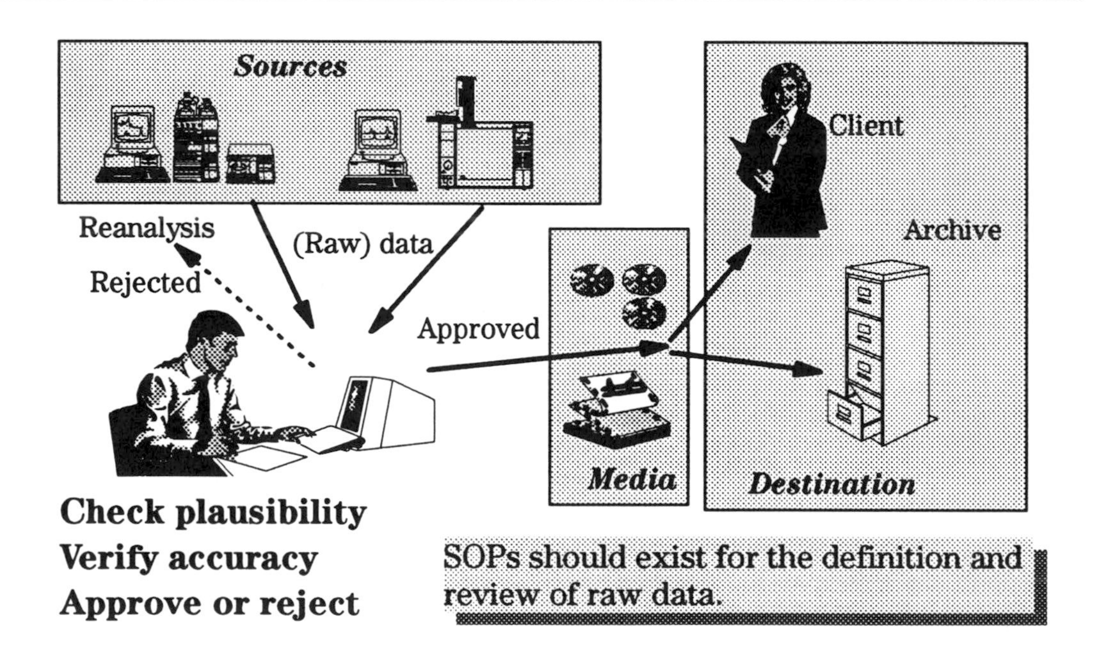

Figure 14.4. Data should be checked for plausibility and approved or rejected. SOPs should be available for the definition and review of raw data.

analytes are expected to be present close to the detection limits and possibly interfering chromatographically with the chemical matrix, there is a high risk of wrong identification, integration and quantitation. In this case it is recommended to visually inspect every chromatogram and to reintegrate, if necessary. A 100% check of chromatograms does not make much sense for well defined samples with only few and well separated peaks and when the expected amounts are far above the limit of quantitation. A good understanding of the analysis task, a good knowledge of the measurement process together with a realistic feeling for anticipated problems, supported by statistical data, are the basis for a good scientific judgment of the extent of data validation.

Security and Integrity of Data and Back-up

Security

If a computer system contains confidential or safety related information, suitable and adequate procedures should be in place to ensure the security of the system. Implementation of security features within a computerized system is also an essential element for establishing and maintaining ongoing control of the system. The pharmaceutical industry and regulatory agencies pay considerable attention to computer system security. For example, the US PMA has addressed this topic through its Computer System Validation Committee in an article on security concepts and the validation of security.[13] The EPA has included a chapter on security of equipment in its GALP draft recommendations[20] and section 10.7 of the ISO/IEC Guide 25[59] recommends; "Procedures shall be established for protecting the integrity of data; such procedures shall include, but not be limited to, integrity of data entry or capture, data storage, data transmission and data processing." The *EURACHEM/WELAC Guidance on the Interpretation of the EN 45000 Series and ISO/IEC Guide 25*[60] also includes several paragraphs on data protection.

As illustrated in Figure 14.5, three major aspects should be considered when establishing a security system:

1. Data must not be lost (prevent information loss through ensured availability).

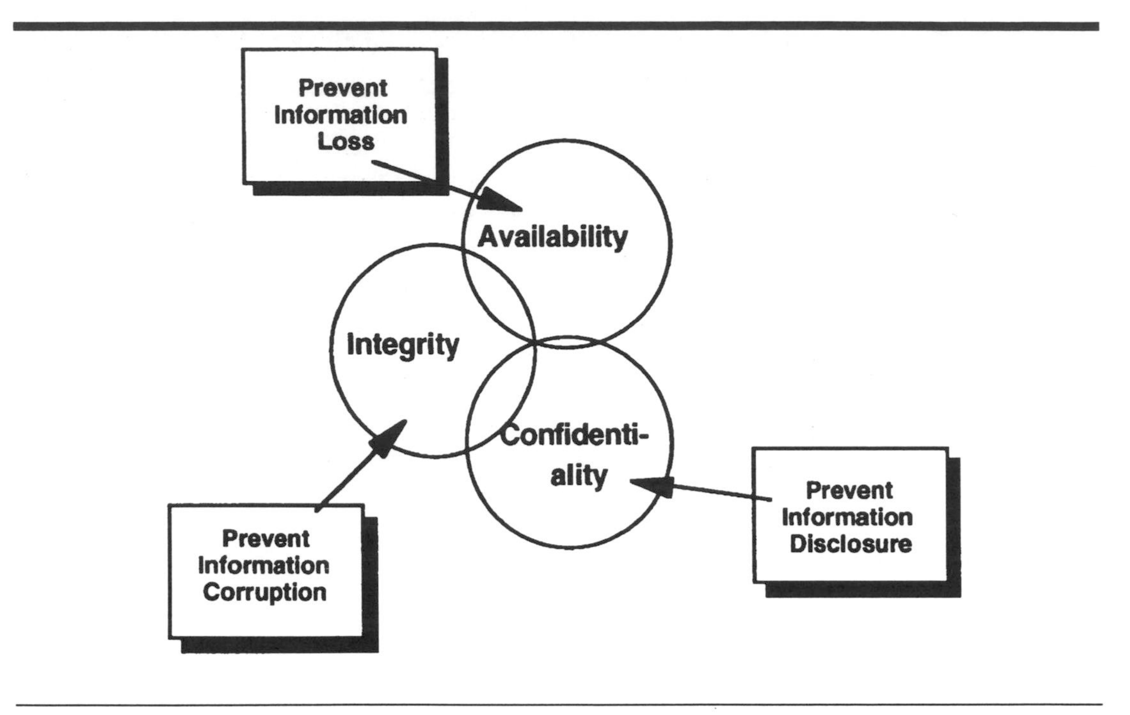

Figure 14.5. Security aspects[20]

2. Data must not be changed either intentionally or accidentally without authorization and documentation (prevent corruption through ensured integrity).
3. Data must not be used for purposes other than planned (prevent disclosure through ensured confidentiality).

Security mechanisms include limiting physical and/or functional access to authorized personnel, accomplished by limiting physical access to the computers (physical security) or through built-in software security systems (logical security) or a combination of both. System access should be established for application software programming and for the configuration and operation of the computerized system. It is often not feasible to place computers that control and evaluate analytical equipment in specially locked computer rooms. In this case, access to the facilities should be controlled and computers should either be secured physically with keyboard locks or through appropriate user identification mechanisms such as passwords. All security strategies should be understood and followed by the operators and should be routinely checked and evaluated.

Establishing a security system requires three key steps:

1. Definition of the requirements with a risk assessment.
2. Design and implementation of the system.
3. Validation of the system.

A risk assessment should be made, for example, on the selection and checking of electronic media and on conditions for archiving. Risks include loss of data through damage caused by a less than ideal environment, such as inadequate temperature, high humidity or electromagnetic fields. Possible efforts to lower the risk of data loss could include selecting archives with ideal conditions, saving the data in duplicate with storage in separate rooms and periodically rereading and copying the data to new media. The costs required for this effort should be balanced against the loss that would arise if data were lost.

Similarly, the risk of equipment failures and subsequent lack of analyses data during repair should be estimated. Efforts to lower the risk could include regular preventive maintenance and substitute equipment. The costs should be compared to the potential loss of being unable to analyze samples for the duration of the repair. A third example is the risk assessment of data loss caused by a power failure.

The exact set of security measures implemented for a specific computerized system is determined by the application as well as by the laboratory's risk assessment. Table 14.2 includes proposed steps that should be considered when building up a security system. It is most important to critically evaluate and implement only the steps that are required to ensure a specific system's security. For example, before developing a contingency plan for emergency operations in the event of total system failure, one should assess the risk of how often this is likely to occur and give careful thought to whether back-up systems are necessary or if one can manage without the system until operation is restored. An overpowered security system increases administration effort and overall costs. On the

Table 14.2. Possible strategies for establishing a security system

Task	Steps
Identify and define requirements and assess risk	**Identify and define requirements** • Identify and define regulations, company policies, client requirements (accessibility of data, availability, corruption, disclosure, integrity). • Define confidential and safety related data. • Define user access to different programs, functions and data. **Assess the risk** • Assess the vulnerability and threats to the system and the consequent potential risk of deliberate or accidental loss or change of data through hardware or software failure, human error, power failure, electromagnetic interference and system access by unauthorized personnel.
Design and implement security systems	**Prevent unauthorized access** • Lock computer rooms with log-in and log-out of all unauthorized personnel accessing the computer room (physical security) • Lock computers or keyboards. • Limit system access through periodically changed passwords and/or passcards (also important when connected to other computers through a network or modem). • Limit program access through passwords (logical security). **Prevent unauthorized installation and use of external software** • Develop, publish and enforce such policies. **Prevent deliberate or accidental change and loss** • Check for viruses. • Back-up and archive software, methods and data in duplicate and segregate copies. • Develop a disaster recovery plan. • Save data on WORM disk drives (Write Once Read Many). • Archive in fireproof, flood resistant, theftproof rooms that are shielded from strong magnetic and electrical fields.

Continued on next page

Continued from previous page

	Identify corrupt or changed data • Periodically reread data before risk of damage. • Implement checksum procedure. **Familiarize operators with the security system**
Validate security features and document results	• Develop an ongoing validation plan that checks all security features. • Implement the validation plan by checking all specified security features. • Document results.

other hand, lost or corrupt data can also be extremely costly; therefore, an appropriately thorough risk analysis is of the utmost importance.

Back-up and Recovery

Documentation should exist to clearly describe the procedures necessary to create and store a back-up copy and for the recovery of system data, including system configuration files. The frequency and type of back-up files should be included in the procedures.

Security Features Built into the Computer System

Modern computer systems have several security features built into the software and hardware. These include:

Limited accessibility

- Cover locks on computer systems prevent unauthorized access to the computer's internal components.
- Keyboard locks
- Password prevents unauthorized use of the computer.
- User specific password I.D. for specific application programs and functions

- Video blanking mode prevents anybody from reading the content of the screen.

Integrity

- Data stored on optical disks with higher safety
- Checksum procedure for the identification of changes to data files
- Binary data file structure makes intentional data manipulation more difficult.
- Programs for automated identification of viruses

Automated error prevention and identification

- Warning messages in case of potential disk overflow when preparing the system for automated analysis and data acquisition
- Recording and reporting of instrument errors and any unusual events for recognition of potential problems
- Instrument shut-down in case of safety problems or potential data loss

Traceability

- Storage of equipment identification, analytical parameters and operators together with raw data files for traceability

For example, the Hewlett-Packard ChemStations have a number of security features built into the software for security, traceability and integrity. Access to the system is restricted to authorized users through password protection in the software. The ChemStation's data handling preserves integrity by storing initial raw measurement data together with instrument conditions and the instrument's logbook in a single checksum protected, binary coded register file.

The HPLC ChemStation records the pre-column pressure and temperature of the column compartment before, during and after the HPLC runs and stores these profiles. In a CE run, the ChemStation stores actual current, voltage and power and

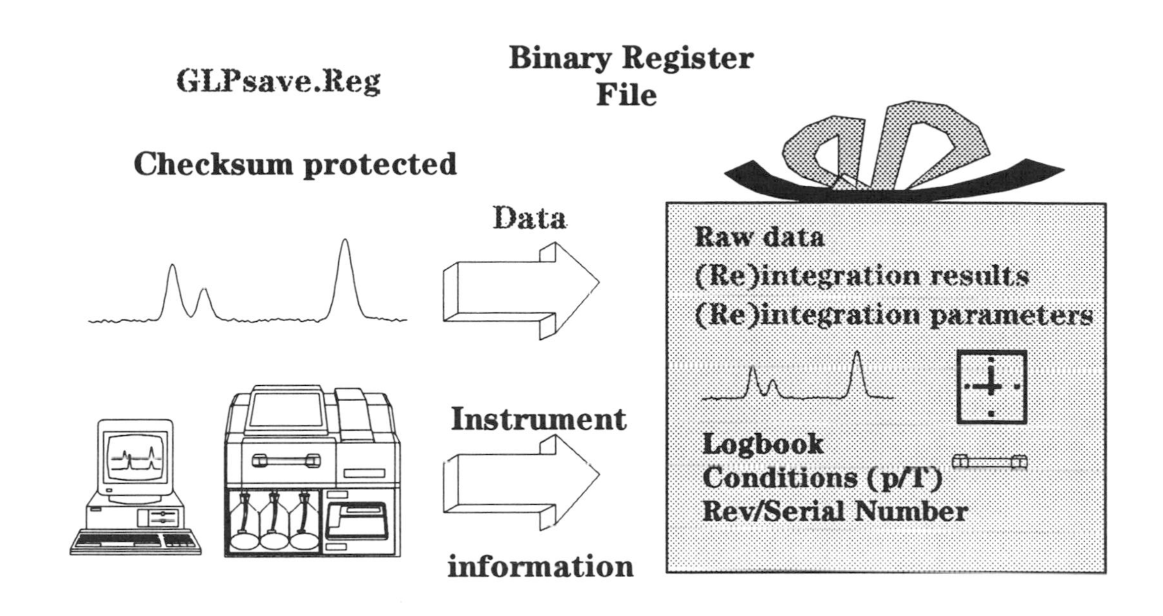

Figure 14.6. In the HP GC, HPLC and CE ChemStations, instrument parameters are stored together with raw data in checksum protected binary register files.

capillary temperature and pressure. This facilitates traceability of data to analytical parameters. Checksum is a programming terminology for an arithmetic operation performed on the data immediately after generation, the product of which is stored with the data. Future access to the data is subject to the same arithmetic check. Numerical matches confirm that data has not been tampered with, while mismatches draw attention to possible data corruption.

15. Diagnostics, Error Recording and Reporting

An analytical system can only produce high quality data when it operates error-free, therefore, the system should be continuously evaluated for proper functioning. Errors should be detected, recorded and their impact on the data assessed and documented. In the worst case when a serious error occurs, the analysis must be re-run. Some instrument errors are easily identified while others are more difficult to identify. For example, a blown main fuse will immediately shut the instrument down but if a character in the display shows a 1 instead of a 7, it may take some time before the error is noticed. The other problem is knowing when the error first occurred. In this case, microprocessor-controlled equipment together with sophisticated computer software has a tremendous advantage. Firmware built into the hardware modules identifies malfunctions, while system malfunctions are detected by software running on the computer system. More intelligent software can also diagnose the source of the problem, estimate the extent to which data is affected and make recommendations on how the problem may be resolved. If the malfunction could cause a safety hazard, for example, a leak in an HPLC system, or if there is a risk of incorrect data being produced, the system should be shut down. Often, such errors can be automatically recorded and stored with a date and time stamp in the system's electronic logbook.

An example of intelligent error handling in a computerized HPLC system is illustrated in Figure 15.1. The system's firmware continuously checks all important functions, and should an error occur, a red LED illuminates and the error is entered into the equipment's electronic logbook. If there is a safety problem (e.g., a solvent leak) the instrument switches off and is also capable of switching off other modules in the system. For

- Microprocessor checks equipment for errors
- Error message on display
- Error recorded in equipment logbook

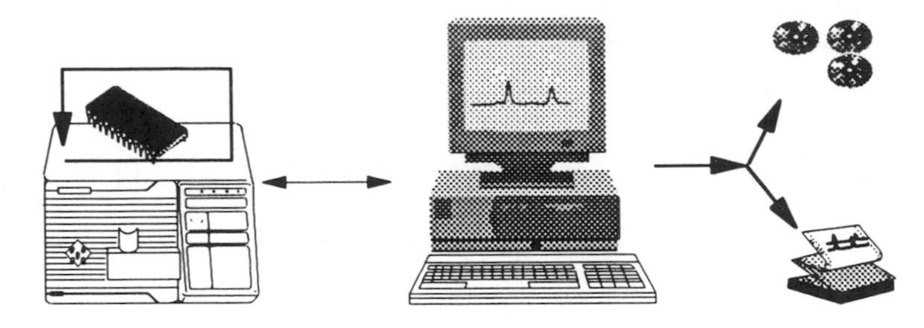

- Errors messages transferred to computer
- Computer initiates actions (for example, terminates a sequence)

- Logbook stored with raw data

SOPs should exist for handling of errors.

Figure 15.1. Error handling in a computerized HPLC system

instance, when a leak is detected in the sampling system, the pump will be switched off to prevent further solvent flow.

The computer checks all components of the system and stores any error messages in the system logbook together with the date and time.

Some tests, the ROM (read only memory) test, for example, are performed every time the instrument is switched on; others such as the leak test are performed routinely, and finally tests like the lamp intensity profile test are performed on request. Table 15.1 lists diagnostic routines employed by an HPLC system. Other analytical systems may use the same or similar tools.

Error information from all HPLC modules, including errors detected in the computer system, are collected and entered into a system logbook. The logbook is saved as a file that can be displayed and printed. The operator can obtain information on the type and source of an error and can ask for recommendations on possible actions to solve the problem. Examples of errors with information on the remedial actions available to the operator are shown in Table 15.2.

Table 15.1. Diagnostic routines in an automated HPLC system

HPLC System

- Read Only Memory (ROM) test
  This test is performed automatically every time the instrument is started. It checks the integrity of the ROM processor by comparing the actual checksum number with the original checksum number burned into the ROM.

- Random Access Memory (RAM) test
  Run during instrument start-up, a series of numbers are written to and read from the processor RAM memory. Both series of numbers must be identical to pass this test.

- Display test
  To ensure that all important user information is visible, the operation of all display devices including LEDs, status and error lamps are checked.

- Remote connections
  This tests the communications to and from external devices and checks their status: ready, not-ready, error. This is an important function that enables any module to shut down the pumping device should a leak be detected anywhere in the system.

- Leak sensors
  The detection of a leak switches off the pumping system to protect operators from a potential health hazard, prevent damage to the instrument from flooding and minimize any potential fire hazard. The functioning of the leak sensors is checked constantly by the system.

Pumping devices

- Metering devices, drive power, pump motor
  The correct operation of the metering devices, the drive power to the pump motor and the status of the ball valves are constantly monitored. A liquimeter measures the volume of solvent pumped. Additionally, the position of the pistons must be within certain predefined limits; any significant deviation from the reference position is displayed.

- Pressure and pressure ripples
  The pressure and any pressure ripple within the system are constantly checked to avoid column damage from fast pressure changes and to immediately detect blockages or low pressure caused by empty solvent reservoirs or gas bubbles. The pressure profile may be plotted on a recorder or integrator.

Continued on next page

Continued from previous page

- Leak tightness
  After loading the leak test method and connecting the pump outlet to a steel blank nut, the pistons are reset to a predefined position. The pump runs several phases with time-programmed flow conditions that should lead to a certain pressure profile. Any deviation from this profile indicates a leak somewhere in the pumping system. The leak-tightness test can be performed by either the system operator or an HP technician.

Automated liquid samplers

- No vial detected
  It is possible to program the instrument's response when no vial is found in the injection position. The operator may choose to run the method anyway without an error message, or to stop the current method and run an alternative "in error" method that either flushes or shuts down the entire system or both.

- Step-functions for servicing
  The step-functions are a valuable tool for servicing the autosampler allowing each step of the injection cycle to be performed individually for verification.

Detectors

- Lamp intensity profile
  The lamp intensity is an important parameter that should be checked regularly. By activating the built-in intensity profile test, the lamp's energy output is measured between 210 and 350 nm and the maximum value is displayed. Tracking these measurements gives an indication of the long-term behavior of the lamp.

- Electronic noise
  Electronic noise is assessed by testing the electronics with the light path blocked so that results are not influenced by the optical unit.

- An ASC-board test measures the dark current of the photodiodes.

- Stray-light is measured to determine whether it is within certain absorbance limits.

- A digital-to-analog converter test checks the digital-to-analog output board by generating a special test pattern.

Table 15.2. Modern computerized systems employ intelligent error diagnosis and recommend actions to solve problems. The error messages are permanently stored in the system's logbook.

Error	Recommended Action
Pumping devices	
Pressure limit exceeds upper limit	• Check that the flow setting is not too high. • Check the fittings, capillaries and column frits for blockage. • If error persists, replace the RAD board.
Initialization failed	• Contact Hewlett-Packard Service.
Pump has detected that outlet valve sticks	• Clean the outlet valve. • Clean or replace the filter in the outlet valve.
Adjustment between first and second piston is out of limit	• Make sure that connector J15 on the HPS board is inserted correctly. • Make sure that the PDC board is inserted correctly. • If the error persists, contact Hewlett-Packard Service.
Pressure ripple exceeds limit	• Check that the solvents are thoroughly degassed and that the compressibility factor is set correctly. • Check the tubing and connections between the solvent reservoir(s) and the pump.
Automated sampler	
No vial in sample tray	• Check the vial for correct positioning. • Make sure that connector J10 (vial sensor) on the ALM board is fully inserted.
Plunger failed to draw sample from vial	• Make sure that connector J19 (metering device) on the ALM board is fully inserted. • Make sure that connector J18 (position sensor) on the ALM board is fully inserted.
Column heater	
Fuse blown	• Install new fuse in the column heater.
Detector	
Low illumination	• Deuterium lamp deteriorated. Exchange lamp.
Deuterium lamp has failed to ignite	• Check the deuterium lamp and lamp onnections. • Replace the power supply.

All logbook entries belonging to a particular analysis are stored together with the raw data file. To get more information on the type and source of the error and possible actions to solve the problem, operators can double click on the error message. Information as shown in Table 15.2 is displayed. The operator can locate the problem source and initiate actions to resolve the problem.

16. Audit/Inspection of Computerized Systems

Internal audits are a key element of any quality system. Their objective is to evaluate activities and existing documentation to determine if they meet predetermined internal and/or external standards and/or regulations or customer requirements. Third party laboratory audits are commonly used to ascertain that a laboratory complies with national or international quality standards such as the ISO 9000 and EN 45000 series or to check if they are competent to perform analyses as specified in contracts with clients. Regulatory agencies inspect laboratories to confirm their compliance with GLP, GCP and GMP regulations.

Because computers have almost completely taken over instrument control, data evaluation and report generation in analytical laboratories, internal quality assurance departments, regulatory agencies and private auditors pay considerable attention to their validation. Quality standards and GCP, GLP and GMP regulations provide very little detail on how analytical instruments should be validated. Guidelines on the inspection of computer-controlled systems have been provided by agencies in different countries as part of general GLP/cGMP inspection guides. Agencies have arranged training courses, and inspectors from the regulatory bodies have given presentations and published papers on what should be inspected and which documents they would expect to find to support the use of a computerized system in a regulated environment.[36,64,74,112,113] For example, South Africa has an extensive inspection checklist as part of its GMP regulations.[114] Module three includes 74 checklist items on electronic data processing.

Recommendations on auditing computer systems can also be found in the literature. Kuzel, a member of the US PMA's Computer System Validation Committee, published an article on "Quality Assurance Auditing of Computer Systems."[115]

The procedures in the article are quite detailed and comprehensive and are recommended as an initial system audit. Grigonis and Wyrik published an article on 'Auditing Computer Systems for Quality'.[116] The article provides guidance on auditing computer systems throughout the system life cycle.

Inspectors from various countries, for example, Trill,[36,113] Guerra[64] and Tetzlaff[112] published papers on aspects of inspecting computer systems in general and computerized laboratory systems in particular as used for process and quality control in Good Manufacturing Practice (GMP), Good Laboratory Practice (GLP) and Good Clinical Practices (GCP) regulated environments.

Trill[36,113] and Tetzlaff[112] gave examples on the documents that should be generated during software and system development and made available for audits. They also gave practical suggestions on how to ensure accuracy and reliability of vendor supplied software.

Trill,[6,113] Tetzlaff[112] and Bruederle[117] reported on the problems and deficiencies most frequently found during inspections of

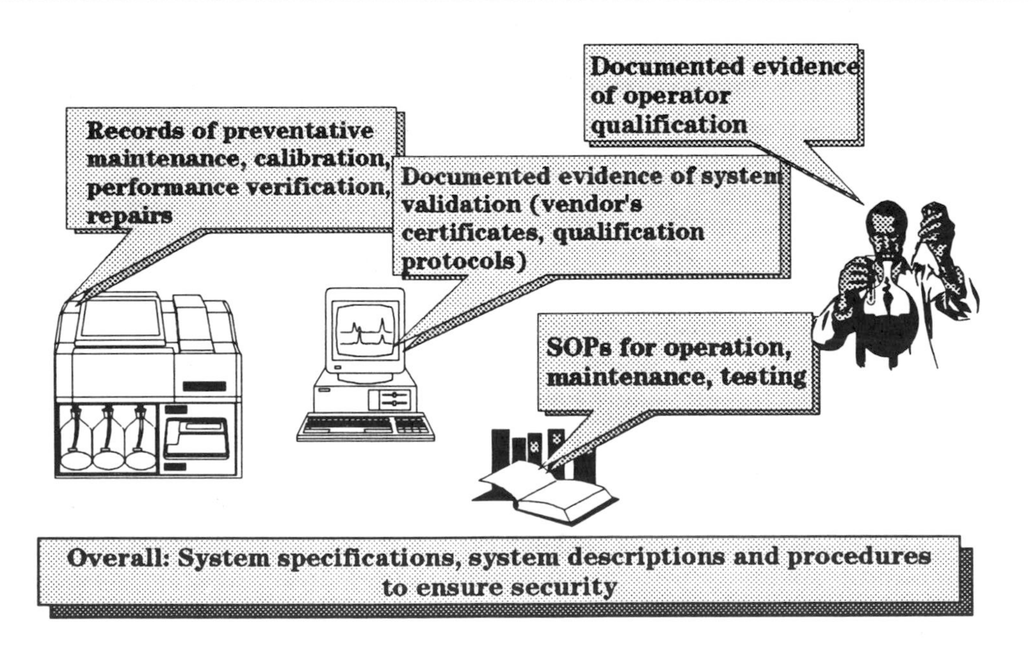

Figure 16.1. Required documentation for system validation

computerized systems and pharmaceutical analytical laboratories. The findings from all three papers have been combined, categorized as far as possible and listed in Table 16.1.

Auditing of computerized systems includes elements similar to manual systems:

- Staff interviews
- Examination of documentation
- Observation of laboratory work

For validation activities including calibration and testing, special attention should be paid to maintenance and change control, safety procedures, back-up and recovery, error handling and recording. Table 16.2 lists possible audit items. Even though there is some doubt about the usefulness of such checklists they do help to ensure that users have considered the most important requirements.

Table 16.1. Problems found during audits of computerized systems in pharmaceutical manufacturing

Life cycle and change control records (for activities at the vendor's site)

- There were no adequate design and performance specifications.
- There was no documentation for the original system from concept to implementation.
- There were no system schematics, data flow or logic diagrams.
- Responsibilities for the system, its documentation, procedures, change control, testing arrangements and data management were not defined.
- There was a lack of procedures for controlling changes, and no detailed records of changes were kept.
- Detailed written descriptions of the computer systems were not kept up to date.
- Original functional design specifications did not provide for future extensions or enhancements of the system.

Software validation and testing

- Software written by the user to customize commercially available software was not subjected to formal documented testing.
- Qualification and validation arrangements for the system were poor with a lack of formal protocols, acceptance criteria, testing procedures, records, reviews, error handling arrangements, formal reporting and signing off.

Continued on next page

Continued from previous page

- The firm had no formal written validation program covering inspections, checks and testing to demonstrate proper performance.
- Validation test logsheets gave an incomplete record without explanation.
- Retrospective validation/evaluation requirements for poorly documented existing systems were not defined or targeted for completion.
- There was no data to assure that the HPLC computer system/software had been validated.

Equipment calibration

- There were no written procedures containing specific directions and/or limits for accuracy concerning the calibration of laboratory instruments and equipment used in finished products testing including HPLC instrumentation, UV-visible spectrophotometer, AA spectrophotometer, ovens etc.
- Written procedures did not ensure that the dissolution apparatus was calibrated each time the shaft was changed for paddles and baskets. The SOP required calibration of the unit only once per year.

Data Security/Integrity

- Disaster/breakdown arrangements did not work/were not validated. Data was corrupted following power loss and shutdown. Poor recovery required data to be rekeyed on restoration of power.
- Memory capacity was limited with overwriting of data on hard disks within several weeks.
- An inability to back-up onto floppy disks was identified.
- There was no program for making back-ups of product specific assay methods.
- There was no system for securing programmed methods.

Software distribution

- The distribution, retrieval and disposition of the software floppy disks was not documented.
- There were multiple versions of the same software program located in various places throughout the facility.

Internal audits

- There were no internal routine system audits by QA.
- Noncompliance reports for process or product parameters were lost in a mass of routine acceptable data in piles of printouts.

Reporting of data

- The final report was undated, unsigned and showed erroneous values.

Continued on next page

Continued from previous page

Method validation/system suitability testing

- There were no written procedures for the validation of analytical procedures and test methods. The suitability of such testing methods was not verified to ensure that they were compatible with conditions that exist in this facility (equipment, environment, personnel, etc.). Written procedures did not specify in-house limits for variable operating parameters that could affect accuracy, reliability and reproducibility of test methods adopted from standard references and compendia.
- HPLC analytical method validation of the two methods reviewed found the following: no determination of precision, specificity, accuracy and no statistical analysis on nonlinearity data.

Analytical quality control

- Management had not established time limits or limitations for the number of HPLC samples/injections that can be run consecutively without reinjecting reference standards.
- System suitability standards and reference standards were not reinjected at the end of lengthy multiple-sample runs to ensure that the integrity and reliability of the procedure was maintained.

Audit trail

- PC-controlled instruments with general purpose operating systems and editing screens enabling changes to be made to analytical procedures and data were not always well controlled.
- Laboratory Information Management Systems (LIMS) did not have forced logging of original method and data (before and after changes) with reasons for change, authorization, identities, times and dates.

Personnel

- There were no detailed user-training records.

Corporate policy

- Corporate policy and controls had not been developed beyond mainframe business systems. Consistent policies, controls and system documentation were not in place for PCs and networks used in production and quality control.

Vendor-user relationship for purchased computerized systems

- There was an absence of formal contracts covering technical, QA and GMP requirements for suppliers providing software and systems and specifying limits of responsibility.

Table 16.2. Audit questions for a computerized system

Software developed in-house

- Do documented procedures for development and testing exist, are they adequate and are they followed?
- Is there a validation protocol, is the content appropriate and did the validation as performed adhere to the protocol?
- Are system requirement specifications available?
- Do system design specifications exist?
- Do module/functional design specifications exist?
- Do system schematics and descriptions exist?
- Is there a description of possible interactions and links with other systems?
- Is there a source code listing?
- Are there module test plans and results?
- Are there system test plans, acceptance criteria and test results?
- Do test data sets represent realistic data?
- Are there documented procedures for change controls?
- Are historical files of changes to programs maintained?

Software and systems purchased from a vendor

- Is there a policy and procedure for purchasing computerized systems?
- Has the vendor been qualified?
- Does the vendor have an established and maintained quality system?
- Does the vendor provide evidence of validation?
- Can validation documents be made available?
- Can the source code be made available to regulatory agencies?
- Is there a software/system tracking and response system for defect reports and enhancement requests?
- Is there a description of what the software or computer system does and of the intended use?

Operation

- Is there a protocol of acceptance testing with test cases, acceptance criteria and test results?
- Do test data sets represent realistic data?
- Have there been manual recalculations of selected jobs?
- Is there a preventive maintenance schedule?
- Is there a schedule for ongoing calibrations and performance verification?
- Is there a record on system calibration, performance verification and maintenance?
- Are instruments labeled to indicated next performance verification/calibration check date?

Continued on next page

Continued from previous page

Method validation

- Has the scope of the method been specified (criteria, performance limits)?
- For standard (compendial) and non-standard methods:
  - — have these methods been validated for all performance criteria as specified by the laboratory and are the results documented?
  - — has the suitability of such methods been verified to ensure that they are compatible with conditions that exist in the laboratory (equipment, environment, people)?
- For nonstandard (noncompendial) methods: is there any documentation showing that these methods are equal to or better than standard (compendial) methods?
- Does a protocol exist for the changes that would cause a revalidation?
- Are methods periodically qualified after the initial validation?

Raw data

- Is there an SOP for defining, collecting, entering, verifying, changing and archiving (raw) data?

Documentation

- Is existing documentation (user manuals, on-line help, SOPs) adequate and up-to-date?
- Is there an equipment logbook?

Audit trail

- Do inputs or changes to data include information on who entered them, and if they were changed, when and why?

Error handling and recording

- Are errors detected and recorded automatically by the system?
- Are there documented procedures on error corrections?

Reporting

- Do reports have adequate information?
- Have reports been dated and signed?

Operator qualification

- Are operators adequately trained for their job?
- Are training records kept?
- Is there an annual review of the training plan?

Reviews

- Is there a documented procedure for inspections or audits?
- Have annual reviews been conducted?

Continued on next page

Continued from previous page

Security

- Is adequate security established to prevent unauthorized access to the system and loss or changing of programs, data or control parameters?
- Is there a list of persons who are authorized to enter and correct data?
- Have security features been validated?
- Is there a documented procedure for back-up of programs and data?
- Is there a documented procedure for disaster recovery?
- Is the extent of any environmental radio frequency known, and are systems protected from the possible adverse effects of radio frequency interference or other electromagnetic interference?
- Is there a documented procedure requiring routine verification that archived software and data remain sound and have not been damaged by physical handling and/or electromagnetic fields?
- Is there a procedure to prevent an incorrect disk from being loaded into the computer?
- Is there a procedure to check all programs for viruses?

Appendix A. Glossary

Acceptance criteria The criteria a software product must meet to successfully complete a test phase or to achieve delivery requirements.

Accreditation The procedure by which an authoritative body gives formal recognition that a body is competent to carry out specific tasks.[118]

Accuracy The degree of agreement of a measured value with the actual expected value.

alpha(α)-test A verification test performed on a newly implemented system that mimics typical operation and that is performed by company personnel who are not system developers.

ANSI American National Standards Institute. Official standards body representing the US with the International Organization for Standardization.

AOAC Association of Official Analytical Chemists.

Application software A program adapted or tailored to the specific requirements of the user for the purpose of data acquisition, data manipulation, data archiving or process control.

AQC Analytical Quality Control.

ASQC American Society for Quality Control.

Audit tracking A procedural formality built into the operation of a system that ensures all interactions with the system are first authorized interactions before being carried out and then recorded permanently in an operations log.

beta(β)-test A verification test performed at a later stage in implementation, after debugging of the alpha-test version and at the customer's site.

Black-box testing A system/software test methodology that is derived from external specifications and requirements of the system.

Methods for black-box testing include random testing, testing at boundary values, and a possible error list. It verifies the end results at the system level, but does not check on how the system is implemented. Nor does it assume that all statements in the program are executed. *See also* functional testing.[24]

BSI British Standards Institution.

Bug A manifestation of an error in software.

Calibration The set of operations that establish, under specified conditions, the relationship between values indicated by a measuring instrument or measuring system, or values represented by material measure and the corresponding values of the measurand. Used by regulatory agencies to refer to the process of checking or adjusting instruments (including analytical instruments). Also used in chromatography to refer to the process of using standard samples as part of method verification.

Change control A procedural formality required for validation, defining how and when changes may be made and in which situations revalidation is required.

CE Capillary Electrophoresis

CEN/CENELEC Comité Européen de Normalisation/Electrotechnical Standardization: The joint European Standards Institution. Develops norms such as EN 45000 series.

Certification Procedure by which a third party gives written assurance that a product, process or service conforms to specified requirements.[118]

Certified reference material (CRM) A reference material, one or more of whose property values are certified by a technically valid procedure, accompanied by or traceable to a certificate or other documentation that is issued by a certification body.[59]

cGMP Current Good Manufacturing Practice.

Checksum Programming terminology for an arithmetic operation performed on the data immediately after being generated, the product of which is stored with the data. Future access to the data is subject to the same arithmetic check. Numerical matches confirm the data has not been tampered with, while mismatches draw attention to possible corruption of data.

Compliance A state of laboratory operations that ensures activities follow documented protocols. GLP compliance is the responsibility of the *study director* who oversees the facility, the personnel, the materials and the equipment or subcontractors that fall under the compliance protocols. A particular instrument is only GLP compliant when *validated* and *verified* by the operator for the specific analysis to be performed. A vendor cannot claim GLP compliance for its products.

Computer system A system composed of computer(s), peripheral equipment such as disks, printers and terminals, and the software necessary to make them operate together (ANSI/IEEE Standard 729-1983).

Computerized system A system that has a computer as a major, integral part. The system is dependent on the computer software to function.[24]

Computer-related system Computerized system plus its operating environment.

Conformity Fulfillment by a product, process or service of specified requirements (EN 45020).[118]

CSVC PMA's Computer system validation committee

Debugging The activity of first determining the exact nature and location of the suspected error within the program and second fixing or repairing the error.

Declaration of conformity A Hewlett-Packard publication that testifies that the equipment has been tested and found to meet shipment release specifications.

Declaration of System Validation A Hewlett-Packard publication that testifies that the ChemStations have been validated during its development and according to the Hewlett-Packard Analytical Products Group Life Cycle.

Diagnostics An early warning system for defects or errors in the system. HP laboratory equipment contains a variety of electronic self-diagnostics that can prevent erroneous measurements being made or flag questionable values in the results report, drawing the operator's attention to nonvalid data.

Disaster recovery plan A document that lists all activities required to restore a system to the conditions that prevailed before the disaster occurred, for example, after power failure.

EC European Community. *See also* EU.

EFTA European Free Trade Association. Members in 1994 are Austria, Finland, Iceland, Liechtenstein, Norway, Sweden and Switzerland.

EN 45001 "General criteria for the operation of testing laboratories." A European standard specifically intended for operating testing and calibration laboratories. EN 45001 is typically used as a guide against which a laboratory's quality system can be evaluated for accreditation. The content of EN 45001 is similar to ISO/IEC Guide 25.

EPA Environmental Protection Agency of the United States Government.

Equipment Defined as the analytical measurement hardware including the firmware, for example, a gas chromatograph. In a computerized system the equipment is controlled by the computer system. The computer system collects measurement data from the equipment.

European Organization for Testing and Certification (EOTC) Formed in 1990 through a Memorandum of Understanding (MOU) between the EC/EFTA members and CEN/CENELEC. Their goal is to harmonize Certification and Accreditation in Western Europe (EC and EFTA) in nonregulated product areas for mutual recognition of standards and accreditation throughout the EC and EFTA countries.

European Union (EU) Members in 1995 are Austria, Belgium, Denmark, Finland, France, Germany, Greece, Ireland, Italy, Luxembourg, the Netherlands, Portugal, Spain, Sweden and the United Kingdom. Formerly called European Community (EC) and European Economic Community (EEC).

External reference specifications (ERS) A Hewlett-Packard document that lists the requirements a new product under development is expected to fulfill.

FDA Food and Drug Administration of the United States Government.

Functional testing Also known as black box testing because source code is not needed. Involves inputting normal and abnormal test cases, then evaluating outputs against those expected. Can apply to computer software or total system.[17]

GALP Good Automated Laboratory Practice.

GAMP Good Automated Manufacturing Practice.

GAP Good Analytical Practice.

GCP Good Clinical Practice.

GMP Good Manufacturing Practice.

IEC International Electrotechnical Commission.

IEEE Institute of Electrical and Electronic Engineers. Technical Committee on Software Engineering.

Interlaboratory test comparisons Organization, performance and evaluation of tests on the same or similar items or materials by two or more laboratories in accordance with predetermined conditions.[118]

International standard Standard that is adopted by an international standardizing/standards organization and made available to the public.[118]

Inspection Structured peer reviews of user requirement specifications, design specifications and documentation.

Installation Qualification (IQ) Documented verification that all key aspects of hardware installation adhere to appropriate codes and approved design intentions and that the recommendations of the manufacturer have been suitably considered.[12,19]

ISO International Organization for Standardization. Agency responsible for developing international standards; represents more than 90 countries.

ISO 9000 series standards The ISO 9000 series standards (9001, 9002, 9003) apply internationally. They are relevant, not just for laboratories, but for all types of manufacturing and service organizations. At this time, the ISO 9000 standards are primarily voluntary. However, many companies are finding that their customers are demanding ISO 9000 series compliance or registration as a condition for doing business.

ISO/IEC Guide 25 "General Requirements for the Competence of Calibration and Testing Laboratories." Like the ISO 9000 series standards, compliance with ISO/IEC Guide 25 is voluntary. Is specifically intended only for calibration and testing laboratories. ISO/IEC Guide 25 is typically used as a guide against which a laboratory's quality system can be evaluated. Unlike the ISO 9000 series standards, it is not possible to apply for an ISO/IEC Guide 25 registration. Content of ISO/IEC Guide 25 is similar to EN 45001.

(Laboratory) accreditation — Formal recognition that a testing laboratory is competent to carry out specific tests or types of tests (EN 45001).[58]

LIMS — Laboratory Information Management System.

LOD — Limit of detection.

LOQ — Limit of quantification.

NAMAS — National Measurement Accreditation Service in the United Kingdom.

National standard — A standard that is adopted by a national standards body and made available to the public.[118]

NDA — New Drug Application. Submission to the US Food and Drug Administration requesting approval to market a new drug. Data must be supplied to support safety and efficacy in humans. Manufacturing and controls and their validation must be described.

NIST — National Institute for Standards and Technology in the United States. Formerly called the National Bureau of Standards Technology (NBS).

Obsolescence — The final phase in a system's life cycle when the system is retired from use and taken off the market. At Hewlett-Packard, an obsolescence plan documents the support activities guaranteed for up to 10 years following obsolescence.

OECD — Organization for Economic Cooperation and Development.

Operational Qualification (OQ) — Documented verification that the equipment related system or subsystem performs as intended throughout representative or anticipated operating ranges.[12,19]

OSHA — Occupational Safety and Health Administration.

Password — In computers and computer software, a security identification text only known to authorized operators, often with capability levels.

Performance Qualification (PQ) — Documented verification that the process and/or the total process related system performs as intended throughout all anticipated operating ranges.[20]

Performance Verification (PV) — A service offered by Hewlett-Packard's Analytical Products Group support organization. It verifies that the system at the user's site performs according to the specifications as agreed between the vendor and the purchaser. Chromatographic performance specifications are published in the vendor's specification sheets. Chromatographic instrument hardware specifications include baseline noise and precision of retention times and peak areas.

PIC Pharmaceutical Inspection Convention, a multinational organization (primarily of European countries) whose members have agreed to mutual recognition of facility inspections for good manufacturing practice.

PICSVF UK pharmaceutical industry computer systems validation forum.

Plug and play The ability to use a system without the need for any written instructions.

PMA Pharmaceutical Manufacturers Association in the United States. A trade association that represents more than 100 firms, collectively producing more than 90 percent of American prescription drugs.

Precision The degree of agreement of a measured value with other values recorded at the same time, or in the same place or on similar instruments. Also referred to as repeatability.

Proficiency testing Determination of laboratory testing performance by means of interlaboratory test comparisons.[118]

Prospective validation Establishing documented evidence that a system does what it purports to do based on a validation plan.[33]

Quality assurance A set of activities, often performed by employees in a similarly named department, that check that the characteristics or qualities of a product actually exist at the time the product is sold.

Records All documents that provide evidence of what you were *going to do,* that you *did* it, and what happened when you had *done* it.

Reference material A material or substance, one or more properties of which are sufficiently well established to be used for calibrating an apparatus, assessing a measurement method or for assigning values to materials.[59]

Reference standard A standard, generally of the highest metrological quality available at a given location, from which measurements made at that location are derived.[59]

Registration A procedure by which a body indicates relevant characteristics of a product, process or service, or particulars of a body or person, in an appropriate, publicly available list.[118]

Retrospective validation Establishing documented evidence that a system does what it purports to do based on review and analysis of historic information.[31]

Revalidation A repetition of validation necessary after the process has been changed, for example, when a manual system is upgraded to an automated system.[33]

Ruggedness An indication of how resistant the process is to typical variations in operation, such as those to be expected when using different analysts, different instruments and different reagent lots. Required under GLP guidelines.

Software subscription An HP service that provides automatic updates of software and documentation.

Source code An original computer program in a legible form (programming language), translated into machine-readable form for execution by the computer.

Standard Operating Procedure (SOP) Documented instructions that should be followed when operating a process for the process to be considered valid. Required under GLP regulations.

STARS A Hewlett-Packard database (Software Tracking and Response System) maintaining historical records of all software revisions, known defects and enhancement requests, a form of change control. Every Hewlett-Packard service center has access to the database over company internal networks.

Structural testing Examining the internal structure of the source code. Includes low-level and high-level code review, path analysis, auditing of programming procedures and standards actually used, inspections for extraneous "dead code", boundary analysis and other techniques. Requires specific computer science and programming expertise.[19]

Study director Person in the laboratory responsible for the outcome of the GLP validation.

Study sponsor Person representing the facility sponsoring an application to the authorities.

System Comprises the analytical equipment and the method used on that equipment. A system of computerized equipment also includes the computer and software (including all mathematical operations performed on the measurement data).

System life cycle — The period of time that starts when a software product is conceived and ends when the product is no longer available for use. The software life cycle typically includes a requirements phase, design phase, implementation phase, test phase, installation and checkout phase and operation and maintenance phase.

System suitability testing — A process of checking out the performance specifications of a system, often called method validation when applied to a particular separation and called system validation when applied to a separation system used routinely.

Test — A technical operation that consists of the determination of one or more characteristics or performance of a given product, material, equipment, organism, physical phenomenon, process or service according to a specified procedure.[59]

Test plan — A document prescribing the approach to be taken for intended testing activities. The plan typically identifies the items to be tested, the testing to be performed, test schedules, personnel requirements, reporting requirements, evaluation criteria and any risks requiring contingency planning.

Traceability — The property of a result of a measurement whereby it can be related to appropriate standards, generally international or national standards, through an unbroken chain of comparisons.[59]

User interface prototyping — A verification test for software performed on design phase subsystems to check that operational concepts are grasped by the operators.

United States Pharmacopeia (USP) — Official pharmacopeia for the United States.

Western European Calibration Cooperation (WECC) — Deals with accreditation of calibration laboratories. Established in 1985 through a Memorandum of Understanding (MOU) for multilateral recognition of accredited laboratories.

Western European Laboratory Accreditation Corporation (WELAC) — Formed in 1989 to represent the interests of laboratory accreditation bodies in Western Europe. Deals with accreditation of test laboratories for European recognition by their clients. Established through a Memorandum of

Understanding (MOU) for multilateral recognition of accredited laboratories. Evaluates accreditation and certification bodies for European recognition of test laboratories accredited by the accreditation body. Also has contacts with the OECD on the relationship between GLP and EN 45001.

Audits accreditation bodies in Europe, for example, the EAM in Switzerland.

Has been combined with the WECC in 1994. Developed and signed a contract with the EOTC for EC wide recognition of accreditation systems (May 13, 1992).

White box testing A software test methodology at the module level in which test cases are derived from the internal structure of the programs. It may execute all the statements or branches in the program to check on how the system is implemented.[24]

Validation Establishing documented evidence that provides a high degree of assurance that a specific process will consistently produce a product meeting its predetermined specifications and quality attributes.[66]

Verification Confirmation by examination and provision of evidence that specified requirements have been met.[59]

WORM Write once, read many. An optical disk drive with huge storage capacity. The disk becomes a read-only storage medium after data is written to the disk.

Worst case 1) A set of conditions encompassing upper and lower processing limits and circumstances, including those within standard operation procedures that pose the greatest chance of process or product failure when compared to ideal conditions. Such conditions do not necessarily induce product or process failure. 2) The highest or lowest boundary value of a given control parameter actually established in a validation qualification exercise.[33]

Appendix B. (Standard) Operating Procedures

Operating Procedures or Standard Operating Procedures (SOPs) play a major role in analytical laboratories' quality systems. This appendix gives some recommendations on (standard) operating procedures, including a list of operating procedures for computerized analytical systems. Examples of SOPs are provided for validation of simple and complex application software, for retrospective evaluation and validation of existing computer systems and for testing of hardware. (SOPs are available in electronic format from the author.)

General recommendations:

- Do not prepare too many SOPs. Think twice before you prepare an SOP for a special job (too much paperwork does not improve the efficiency of a laboratory or the quality of analytical data).
- Where possible, combine individual procedures into a single, larger SOP.
- Do not be too detailed, this avoids the need for frequent updates.
- Instrument SOPs should be written close to the instrument in the laboratory and not in the office. They should be either written or thoroughly reviewed by the instrument's operator. SOPs should not explain how procedures are supposed to work, but how they work in reality.
- Copies of equipment SOPs should be located close to the instruments for easy access by operators.

SOPs could be developed for:

☑ Software development and validation (life cycle)

☑ Retrospective evaluation and validation of existing computer systems

☑ Revalidation

☑ Installation

☑ Routine inspection, testing, maintenance and calibration

☑ Maintaining security

☑ Entry of data and proper identification of individuals entering the data

☑ Actions to be taken in response to equipment failure

☑ Definition of raw data

☑ Changes to data and methods

☑ Analysis, processing, changing, reporting, storage and retrieval of data

☑ Operator training

☑ Development and handling (distribution, archiving, etc.) of SOPs

Example #1: Development and Validation of Simple Application Software

The following is an example of an operating procedure used for the development and validation of a simple application software program developed in a user's laboratory, in this case for the further customization of a commercial standard chromatography software. This is a proposal and starting point only. The amount of validation and documentation depends on the complexity of the program, and the proposed procedure should be adapted accordingly. There is no assurance expressed that this operating procedure will pass a regulatory inspection.

<table>
<tr><td>Company</td><td colspan="2">Title: Development and Validation of a MACRO Program</td><td colspan="2">Code
LHSW01</td><td>Revision number
A.01</td></tr>
<tr><td>Division</td><td>Laboratory</td><td colspan="2">Manager</td><td colspan="2">Page 1 of 3</td></tr>
<tr><td>Effective Date</td><td>Prepared by

Name:
Sig:
Date:</td><td colspan="2">Approved by

Name:
Sig:
Date:</td><td colspan="2">Distribution

—
—
—
—
—</td></tr>
</table>

1. Scope

Validation of application software (e.g. MACROs), written for the Hewlett-Packard ChemStations or other computer systems.

2. Purpose

The purpose of this operating procedure is to ensure that application software (e.g., MACRO programs) is properly validated during its development and periodically reevaluated during its operation.

<table>
<tr><td>Company</td><td colspan="2">Title: Development and Validation of a MACRO Program</td><td>Code
LHSW01</td><td>Revision number
A.01</td></tr>
<tr><td>Division</td><td>Laboratory</td><td colspan="2">Manager</td><td>Page 2 of 3</td></tr>
</table>

3. The validation procedure

a) **Responsibilities**
Identify persons responsible for development, tests, and approvals.

b) **Requirement specifications**
Describe the task and the system (hardware, system software, standard software) requirements.

c) **Functional description**
Describe the program in terms of the functions it will perform, written in such a way that it is understood by both the software developer and the user.

d) **Design and implementation**
Document formulae or algorithms used within the program. Write the code. Document the program such that it can be understood by other people whose education and experience are similar to the programmer. Print the program.

e) **Testing**
Develop test cases and test data sets with known inputs and outputs. Describe the test environment and the execution of tests. Include test cases with normal data across the operating range, boundary testing and unusual cases (wrong inputs). Include procedures to verify calculations. Test procedure and results should be documented, reviewed and approved by the programmer's, and quality assurance departments.

f) **Ongoing performance checks**
Specify type and frequency of checks as well as expected results and acceptance criteria. Develop test data sets for ongoing performance checks.

Company	Title: Development and Validation of a MACRO Program	Code LHSW01	Revision number A.01
Division	Laboratory	Manager	Page 3 of 3

4. User documentation

Describe the program's functionality, formulae used for calculations and how to install and operate the program.

5. Security

Describe which features are implemented to meet security requirements, for example, back-up procedures and limited authorized system access to the program and source code.

6. Change and version control (only required if such an SOP is not already available)

Develop a procedure to authorize, test, document and approve any changes to the software before implementation.

Develop a procedure for clear identification of each software and any revision thereof by program and revision name or code. Develop and maintain a historical file of changes and version numbers.

Example #2: Development and Validation of a Complex Application Software

The following is an example of an operating procedure used for the development and validation of a complex application software program developed in a user's laboratory. This is a proposal and starting point only. The amount of validation and documentation depends on the complexity of the program, and the proposed procedure should be adapted accordingly. There is no assurance expressed that this operating procedure will pass a regulatory inspection.

Company	**Title: Development and Validation of Application Software**		**Code** LHSW02	**Revision number** A.01
Division	**Laboratory**	**Manager**		**Page 1 of 4**
Effective Date	**Prepared by** Name: Sig: Date:	**Approved by** Name: Sig: Date:		**Distribution** — — — — —

1. Scope

Validation of application software by an end-user.

2. Purpose

Quality standards, regulatory agencies and some company policies require software used for evaluation of critical data to be properly validated. The purpose of this operating procedure is to ensure that application software is validated during its development and periodically reevaluated during its operation.

Company	Title: Development and Validation of Application Software		Code LHSW02	Revision number A.01
Division	Laboratory	Manager		Page 2 of 4

3. The validation procedure

Any step can be passed over as long as there is sufficient explanation that the skipped step has no relevance for the program.

a) **Responsibilities**
Identify persons responsible for design, development and testing.

b) **Requirement specifications**
Describe the task, how the problem is solved now and how the new program will solve it more efficiently. Describe user, system, regulatory (GLP/GMP/GCP) and security requirements.

c) **Functional description**
Describe the program in terms of the functions it will perform, written in such a way that it is understood by both the software developer and the user. Review the functional description against the requirement specifications. References may be given to the user documentation.

d) **Design specification**
Define how specified functional and security specifications can best be implemented. Document formulae or algorithms used within the program for data analysis, processing, conversion or other manipulations. Discuss the program design with a second competent person. Review the design specifications against the functional description.

e) **Implementation**
Write the code. Document the code such that it can be understood by other people whose education and experience are similar to the programmer. Decide whether or not to perform a formal code inspection of individual modules. Perform and document a code inspection or give an explanation if there is no formal code inspection.

Company	Title: Development and Validation of Application Software		Code LHSW02	Revision number A.01
Division	Laboratory	Manager		Page 3 of 4

f) **Testing**
Include structural code testing as well as security and functional testing. Verification of structural testing should be performed by providing a matrix that references the source code to the design specification. Develop a test plan for functional testing with the purpose of the test, the functions to be tested, the test steps or methodology and the expected results and acceptance criteria. For functional test, develop test cases and test data sets with known inputs and outputs, describe the test environment, the execution of tests and note how many people should test and for how long. Develop test cases and data test sets that can be used for current and future testing that simulate as much as possible the real-life environment. Include test cases with normal data across the operating range, boundary testing and unusual cases (wrong inputs). Include procedures to verify calculations. Include anticipated users of the software in the test program and perform part or all of the tests in a typical user's environment. Specify how errors found will be classified, documented and what corrective action will be taken. Specify release criteria before the test starts. Test plans and results should be documented, reviewed and approved by the programmer's, user's and quality assurance departments.

g) **Ongoing performance checks**
Specify type and frequency of checks as well as expected results and acceptance criteria. Develop test data sets for ongoing performance checks.

h) **Error tracking system and response system**
Develop a formal feedback system to report any problems and requests for enhancements to the programmer of the software. A team consisting of user(s) and programmer(s) should document, evaluate and classify the problem or enhancement proposal and make proposals for a solution.

Company	Title: Development and Validation of Application Software		Code LHSW02	Revision number A.01
Division	Laboratory	Manager		Page 4 of 4

i) **Change and version control**
Develop a procedure to authorize, test, document and approve any changes to the software before implementation. Develop a procedure for clear identification of each software and any revision thereof by product and revision name or code. Develop and maintain a historical file of changes and version numbers.

4. User documentation and training

a) The user documentation should describe the program's functionality, formulae used for calculations and how to operate the program.
b) Describe which security features are implemented to meet the specified security requirements, for example, back-up procedures and limited authorized system access to the program and source code.
c) Specify the educational, experience and training requirements for the operators of the program.

5. Document storage and archiving

a) Specify what should be located where to ensure easy access by the operator during operation of the software.
b) Specify which documents should be archived.

6. Approvals

a) Approval of the validation protocol by the programmer's, user's and quality assurance departments.
b) Approval of the requirement specification document, design specification document and the test plan by the programmer's, user's and quality assurance departments.
c) Approval and authorization of any changes to the software by the programmer's, user's and quality assurance departments.

Example #3: Testing Precision of Peak Retention Times and Areas of an HPLC System

The following is an example of an operating procedure for the testing of an HP 1050 Series HPLC system for precision of peak areas and retention times. This is a proposal and starting point only and may need adaptation to different HPLC systems. There is no assurance expressed that the operating procedure will pass a regulatory inspection.

Company	**Title: Testing precision of peak areas and retention times of an HPLC system**	**Code** LHEQ01	**Revision number** A.01
Division	**Laboratory**	**Manager**	**Page 1 of 6**
Effective Date	**Prepared by** Name: Sig: Date:	**Approved by** Name: Sig: Date:	**Distribution** — — — — —

1. Scope

Testing the precision of peak areas and retention times of an HP 1050 Series HPLC system.

2. Purpose

The precision of peak areas and retention times are important characteristics for qualitative and quantitative measurements in HPLC. This operating procedure provides chromatographic conditions and key sequences to verify these characteristics of a complete HPLC system, comprising an HP 1050 Series Autosampler, a Quaternary Pump and a Variable Wavelength Detector.

<table>
<tr><td>Company</td><td colspan="2">Title: Testing precision of peak areas and retention times of an HPLC system</td><td>Code
LHEQ01</td><td>Revision number
A.01</td></tr>
<tr><td>Division</td><td>Laboratory</td><td colspan="2">Manager</td><td>Page 2 of 6</td></tr>
</table>

3. Frequency

The precision should be verified at least once a year or after the repair of one or more modules.

4. Instrumentation

a) Quaternary HP 1050 Series Pump
b) HP 1050 Series Autosampler
c) HP 1050 Variable Wavelength Detector
d) HP 3396 Integrator

5. Columns, chemicals

a) Column: 100 mm × 4.6 mm Hypersil ODS, 5μ (HP P/N 79916OD-554).
b) Solvents: Water and Methanol, HPLC grade.
c) Sample: Isocratic standard sample (Hewlett-Packard part number 01080-68704) that contains 0.15 wt.% dimethylphthalate, 0.15 wt.% diethylphthalate, 0.03 wt.% biphenyl, 0.03 wt.% o-terphenyl dissolved in methanol.

6. Preparation of the Variable Wavelength Detector

a) Switch lamp ON.
b) Set the wavelength to 254 nm.
c) Set the response time to 1 SEC.

Company	Title: Testing precision of peak areas and retention times of an HPLC system		Code LHEQ01	Revision number A.01
Division	Laboratory	Manager		Page 3 of 6

7. Preparation of the Pump

a) Prime the pump (use appropriate 1050 SOP): "Priming a Quaternary Pump."
b) Fill solvent reservoirs: A with water, B with water, C with methanol.
c) Degas solvents (see appropriate 1050 SOP): "Priming a Quaternary Pump."
d) Set UPPER LIMIT to 400 (bar).
e) Set the FLOW rate to 3.00 ml/min.
f) Set the temperature of the column oven to 45°C.
g) Set the solvent composition: A = off, B = 15%, C = 70% (channel A will be changed automatically according to %B and %C settings.
h) Set the STOP TIME to 5.00 minutes.
i) Switch pump ON.

8. Preparation of the Autosampler

a) Make sure that the air pressure needed for the solenoid valves is about 5 bar.
b) Switch the autosampler on.
c) Put sample vial with isocratic sample into the vial tray, position number 10.
d) Set up vial numbers: FIRST 10 LAST 10.
e) Set the number of injections/vial to 6.
f) Set the injection volume to 10 μl.

9. Set parameters for the HP 3396 Integrator

a) Attenuation: 10.
b) Chart speed: 1 cm/min.
c) Zero: 10.
d) Threshold: 10.

<table>
<tr><td>Company</td><td>Title: Testing precision of peak areas and retention times of an HPLC system</td><td colspan="2">Code

LHEQ01</td><td>Revision number
A.01</td></tr>
<tr><td>Division</td><td>Laboratory</td><td colspan="2">Manager</td><td>Page 4 of 6</td></tr>
</table>

10. Analysis of Isocratic Standard

a) When the baseline is stable, start thc analyses.
b) As a result, 6 chromatograms similar to the Figure below should be obtained (differences in retention times and areas due to variations between different column batches and to variations in the concentration of the sample from batch to batch).

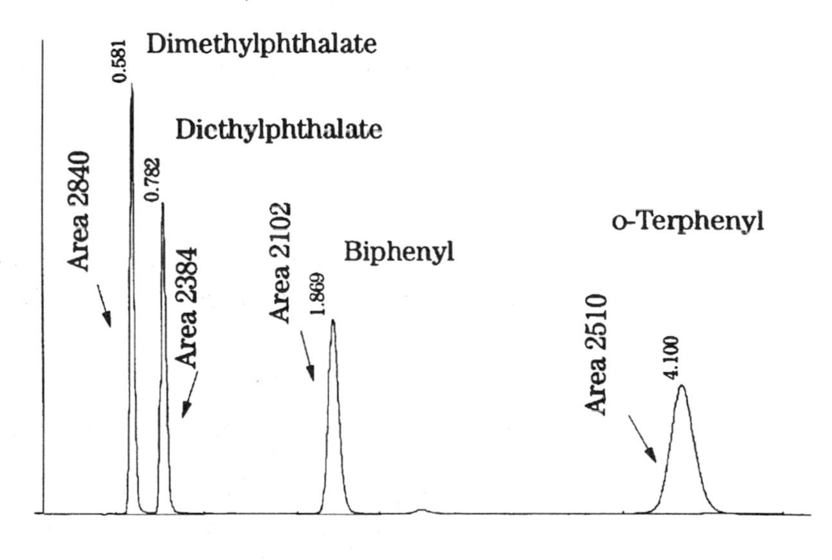

Company	Title: Testing precision of peak areas and retention times of an HPLC system		Code LHEQ01	Revision number A.01
Division	Laboratory	Manager		Page 5 of 6

11. Acceptance

a) Calculate the precision of retention times and peak areas.

$$\text{RSD} = \frac{\sqrt{\frac{1}{n-1}\Sigma(x-\bar{x})^2}}{\bar{x}}$$

where:
n is the number of injections
x is area or retention time of peak

$$\text{Mean} = \bar{x} = \frac{1}{n}\Sigma x$$

b) The precision for the peak areas should be < 1.5% RSD.
c) The precision for retention times should be < 0.5% RSD.

12. Further Action

If the 1050 HPLC system does not fulfill the given specification, do the following:

a) Check the performance of the detector (noise and drift) using appropriate SOP: "Checking Noise and Drift on the HP 1050 Series Variable Wavelength Detector."
b) Check whether the pump is leak-tight using appropriate SOP: "Leak Test for the Quaternary HP 1050 Series Pump."
c) Check whether the autosampler is leak-tight using appropriate SOP: "Checking the Pressure Tightness on HP 1050 Series Autosampler."

If following these procedures does not result in an improvement, call Hewlett-Packard service.

Company	Title: Testing precision of peak areas and retention times of an HPLC system		Code LHEQ01	Revision number A.01
Division	Laboratory	Manager		Page 6 of 6

13. Appendix—Protocol example for results

Instrument identification

Serial number pump: ____________

Serial number autosampler: ____________

Serial number detector: ____________

Serial number integrator: ____________

Date: ____________

Results

Precision of peak areas: ____________ (spec <1.5% RSD)

Precision of retention times: ____________ (spec <0.5% RSD)

Comment:

Further actions (in case the equipment is out of specification)

Approvals

	Name	Signature	Date
Laboratory supervisor	____________	____________	_______
Test engineer	____________	____________	_______

Example #4: Retrospective Evaluation and Validation of Existing Computerized Analytical Systems

The following is an example of an operating procedure for the retrospective evaluation and validation of existing computerized analytical systems. This is a proposal and starting point only and may need adaptation to different systems. There is no assurance expressed that the operating procedure will pass a regulatory inspection.

Company	**Title: Retrospective Evaluation and Validation of Computerized Analytical Systems**	**Code** LHSW03	**Revision number** A.01
Division	**Laboratory**	**Manager**	**Page 1 of 7**
Effective Date	**Prepared by** Name: Sig: Date:	**Approved by** Name: Sig: Date:	**Distribution** — — — — —

1. Scope

Evaluation of existing computerized analytical systems retrospectively for past and current use and prospective validation for future use. The procedure is limited to systems purchased from a vendor.

2. Purpose

Regulatory agencies require computerized analytical systems used for the analysis and evaluation of critical data to be validated. Existing computerized systems in laboratories frequently have not been formally validated or their initial

Company	Title: Retrospective Evaluation and Validation of Computerized Analytical Systems	Code LHSW03	Revision number A.01
Division	Laboratory	Manager	Page 2 of 7

validation was not documented. The purpose of this operating procedure is to demonstrate whether such systems were operating as intended in the past, whether they are currently operating as intended and whether they will operate as intended in the future.

3. Develop a validation plan

Define validation requirements. Define current system expectations, evaluate what was done in the past and what is planned for the future to meet these expectations. For content and details of the plan follow steps 4 to 7 of this operating procedure.

4. Describe and define the system

a) Describe the purpose of the system.
b) List the equipment hardware.
   — in-house identification number
   — merchandising number or name
   — manufacture's name, address and phone number
   — hardware serial number, firmware revision number
   — date received in the laboratory, date placed in service
   — location
c) List all computer hardware.
   — manufacturer's name
   — model, serial number
   — processor, coprocessor
   — memory (RAM), graphics adapter
   — hard disk
   — interfaces, network

Company	Title: Retrospective Evaluation and Validation of Computerized Analytical Systems		Code LHSW03	Revision number A.01
Division	Laboratory	Manager		Page 3 of 7

d) List all software loaded on the computer software with product number, version number and the name of the vendor
   — operating system, user interface
   — canned standard software
   — user specific application software, e.g., MACROs, with date and size
e) List accessories such as cables, spare parts, etc.
f) Find and review or develop system drawings.
g) Define operator requirements.
h) Define all required functions and operational limits of the modules and system as used for the current application.
   — for equipment hardware
   — for the software and for system functions
i) Define physical and logical security requirements, e.g., physical or password access.

5. Collect any documentation available

a) Reports from internal users on number and type of problems.
b) Reports from external users on number and type of problems.
c) Purchase orders.
d) Certificates and specifications from the vendor.
e) Information on what formulae are used for calculations.
f) Operating procedures, for example, for basic operation, maintenance, calibration and testing of the system.
g) User manuals.

Company	Title: Retrospective Evaluation and Validation of Computerized Analytical Systems		Code LHSW03	Revision number A.01
Division	Laboratory	Manager		Page 4 of 7

6. Collect information on system history

a) Installation reports.
b) Information on acceptance testing.
c) System failure reports.
d) Equipment hardware and system maintenance logs.
e) Maintenance records.
f) Calibration records.
g) Results of module and system performance checks.
h) Any test reports.
i) Records on operator qualifications.

7. Evaluate past and current system performance and document results

Evaluate information and documentation collected under items 5 and 6.

a) Check if documentation as collected under 5f and 5g is complete and up to date; for example, does the revision of the existing user manual comply with firmware and software revision numbers?
b) Check if there is evidence of software development validation. Qualification criteria are availability of type and number of documents listed under 5d.
c) Check if the equipment (hardware) has been qualified for proper and up-to-date functions over the anticipated operating ranges as specified in 4h. Generate a matrix with equipment functions as defined in 4h versus results of calibrations and performance checks as defined in 4h.

Company	Title: Retrospective Evaluation and Validation of Computerized Analytical Systems	Code LHSW03	Revision number A.01
Division	Laboratory	Manager	Page 5 of 7

d) Check if the computer system has been qualified for proper and up-to-date functions over the anticipated operating ranges as specified in 4h. Generate a matrix with system functions as defined in 4h versus results of acceptance testing. Check if calculations made by the computer software have been verified.
e) Check if the computerized system is suitable for its intended use as specified in 4h. Generate a matrix with performance requirements as defined in 4h versus results of system tests.
f) Check if the system is secure enough to meet the security requirement specifications as specified in 4i. Check also if the security features have been verified sufficiently.
g) Check if the number and type of errors reported under 6c indicate continuous functioning of the system.
h) Check if the operators were/are qualified for their jobs.
i) Prepare an evaluation report. Make a statement on past and current validation status; whether the system is formally validated (if not define what changes to the system are needed); and make proposals for further validation steps for future use of the system.

8. Prospective validation for future use

a) Update or develop system description, user requirement specifications, operating ranges, user manuals, appropriate SOPs and safety procedures as necessary.

Company	Title: Retrospective Evaluation and Validation of Computerized Analytical Systems	Code LHSW03	Revision number A.01
Division	Laboratory	Manager	Page 6 of 7

b) Update or develop and implement a test and verification plan for the equipment.
   The plan should be developed to verify the performance of the various equipment parameters over the anticipated operating ranges and should include documented test procedures, expected results and acceptance criteria. After the test phase a formal report that documents the results should be generated.

c) Update, develop and implement an acceptance test plan for the computer system.
   Develop a test plan to exercise the various functions of the computer system. Specify the functions to be tested, the purpose of the individual tests, the test steps or methodology, the expected results and the acceptance criteria. Develop test cases and test data sets with known inputs and outputs for functional test. After the test phase a formal report that documents the results should be generated.

d) Update or develop and implement a preventive maintenance plan.

e) Update or develop and implement a calibration schedule and/or a performance verification schedule.

f) Update or develop and implement a procedure for annual system review.

g) Update or develop and implement an error recording, reporting and remedial action plan.

<table>
<tr><td>Company</td><td colspan="2">Title: Retrospective Evaluation and Validation of Computerized Analytical Systems</td><td colspan="2">Code
LHSW03</td><td>Revision number
A.01</td></tr>
<tr><td>Division</td><td>Laboratory</td><td colspan="2">Manager</td><td colspan="2">Page 7 of 7</td></tr>
</table>

9. Approvals

The validation plan, the system definition, the results of the past and current evaluation, the prospective validation plan and the test plans and results should be approved and signed by the user's and quality assurance departments.

10. References

a) N.R. Kuzel, Fundamentals of computer system validation and documentation in the pharmaceutical industry, *Pharm. Tech.*, Sept. 1985, 60–76.

b) J. Agalloco, Validation of existing computer systems, *Pharm. Tech.*, Jan. 1987, 38–40.

c) H. Hambloch, Existing computer systems: A practical approach to retrospective evaluation, in *Computer validation practices,* Buffalo Grove, IL, Interpharm, ISBN 0-935184-5-4, 93–112, 1994.

d) L. Huber, *Validation of computerized analytical systems,* Buffalo Grove, IL, Interpharm, May 1995.

Appendix C. Strategy for Chronological Selection and Validation of Computerized Analytical Systems

Step	Explanation	Examples
Select type of equipment and method	• Criteria: compounds, matrix, detection limit, precision, selectivity, accuracy, concentration range, qualitative or quantitative, throughput, purchase and operating costs, training and space requirements, skill requirements, regulatory requirements, compatibility with existing equipment	Analysis of phenoxyacid herbicides in drinking water, detection limit: 0.01 $\mu g/\ell$, qualitative and quantitative information, 30 samples/day. Method and equipment: solid phase extraction and HPLC/DAD
Define required functions and intended operational limits	• Define intended equipment functions and operational limits • Define intended software functions • Define intended system functions and operational limits	HPLC: binary gradient, flow rate range 0.2 to 2 ml/min, diode-array detector with 10 mm pathlength, baseline noise <= 4×10^{-5} AU, integration, quantitation, peak purity check, interactive and automated spectral library search, qualitative and quantitative report, automated sequencing

Continued on next page

Continued from previous page

Select and qualify vendor	• Develop criteria for vendor selection	Quality system, availability of validation documentation, local support and response time, reputation and experience
	• Check if criteria are met	Through documentation available from the vendor, 3rd party audit, or direct vendor audit
Select and purchase appropriate equipment and options	• Automated gradient HPLC system with UV-visible diode-array detector • Computer and software for instrument control and data evaluation	HPLC system from Hewlett-Packard consisting of the HP 1050 Series LC system and HPLC ChemStation
Qualify modules and systems prior to routine use	• Preinstallation	Get manufacturers recommendations for environmental conditions.
	• Installation	Check if shipment complies with purchase order.
	• Operation	Test equipment, for example, test the precision of amounts and retention times.
Train operator	• Instrumentation technique • Analysis method familiarization by the	Courses, seminars, on-the-job training, equipment vendor, computer-based training
Develop and maintain equipment specific documentation	• Logbook • Equipment binder • SOPs	SOPs for routine maintenance, calibration, testing

Continued on next page

Continued from previous page

Validate methods	• Specify validation parameters and acceptance limits. • Define validation experiments. • Validate and document.	Limit of detection, limit of quantitation, selectivity, linearity, precision, accuracy, ruggedness
Assure ongoing performance	• Develop and implement schedules and procedures for periodic preventive maintenance, calibration and performance verification.	Exchange of UV-visible detector lamps Cleaning of flow cell Calibration of UV grating for wavelength accuracy
	• Develop and implement procedures for error detection, recording and handling. • Develop and implement procedures for change control	ROM check at system boot up Automated shutdown of pump if leak is detected in an autosampler or detector Authorize changes to user written software
	• Develop and implement security relevant procedures.	Limited system access through user specific passwords Data file integrity through checksum routines

Appendix D. Testing of Selected Equipment

This appendix includes recommendations for testing of equipment frequently used in analytical laboratories. The number of instruments included in this appendix will be expanded in future revisions of this book.

There are no detailed recommendations about time intervals. Tests should help to verify that the system is suitable for its intended use. The indented use should be defined by the user and will influence the type of tests, the acceptance criteria and test intervals. Tests and corrective actions should be done sooner than the system is expected to drift out of the criteria. An initial choice of test interval may be made on the basis of previous knowledge, on manufacturer's recommendations and on intuition. The optimal time interval can be found out by initially doing the tests more frequently and to increase the time interval after the system passed the tests several times. The type and frequency of tests should be specified for each system and the results of each test should be documented.

If a computer system is used for instrument control and data evaluation, the functions of the software executed during the tests should be listed. Successful completion of the test sequence can be used as one evidence that the software is fit for its purpose.

Many of the tests can be automated with commercially available software, for example, with the HPLC, GC and CE ChemStations from Hewlett-Packard.

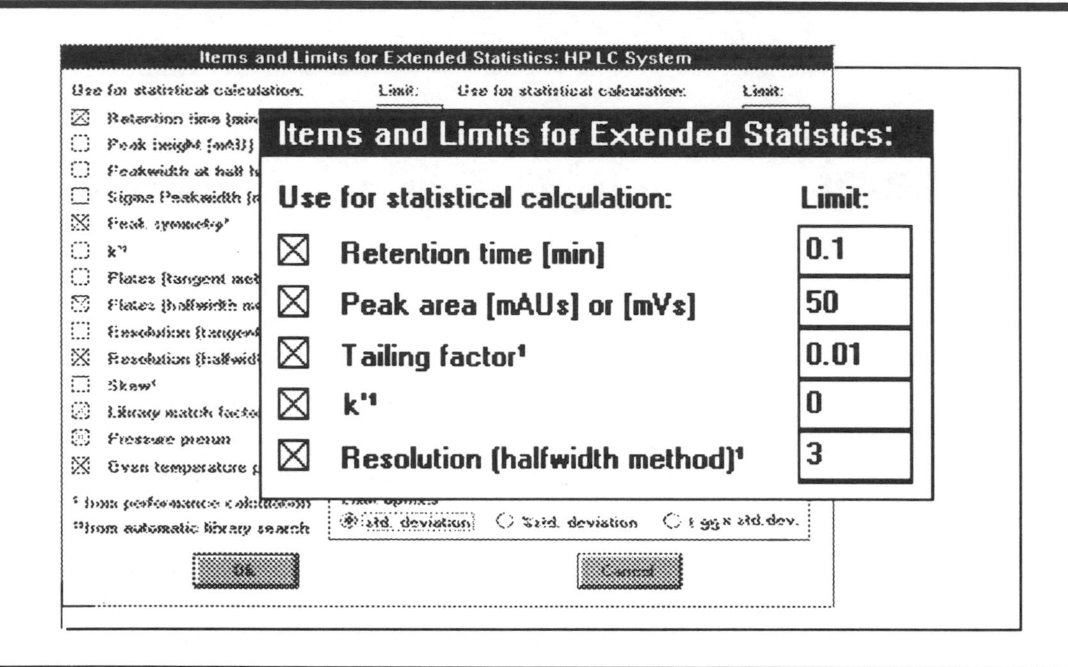

Figure D.1. Commercial software is available from instrument vendors for automated equipment testing. Parameters to be tested and performance limits are selected from a menu.

High-Performance Liquid Chromatography

A complete HPLC system consists of an injection system, a solvent delivery system, a detector and may also include a column thermostat and a computer system. Important HPLC characteristics are the precision of peak areas, retention times, linearity and limits of detection and quantitation. It is recommended to test for these characteristics with a complete system (holistic testing) and not module by module. Tests of individual modules should be conducted if the system does not meet the acceptance criteria.

Quantitation in chromatography and qualification are achieved through comparison of well characterized standards with the unknown samples. Therefore the precision between calibration runs of most performance characteristics such as injection volume, the pump's flow rate and the column oven temperature and the detector response is more important than the accuracy of these parameters. For example, there is hardly any scientific reason to attempt to measure the accuracy of the injection volume.

If methods are transferred between different instruments, the accuracy of some of the parameters will be important. For example, the column temperature, the flow rate and the gradient composition can have an influence on the absolute and/or relative retention times. Testing of flow rate accuracy and gradient composition may also be useful to diagnose retention time drifts in isocratic and gradient operation respectively.

It is recommended to degas the mobile phase(s) thoroughly prior to testing. The system should be purged with solvent and then equilibrated until a stable baseline is obtained.

Precision of Retention Times and Peak Areas

Precision of retention times and peak areas are important characteristics that influence qualitative and quantitative results. A standard is injected five or six times and the absolute and relative standard deviations of peak areas and retention times are measured. The standard should be stable and a column known to have previously performed well should be used. This should effectively eliminate variability due to integration and column performance.

Baseline Noise of Systems with UV-visible Detectors

Baseline noise is an important characteristic that influences the limits of detection and quantitation. Factors contributing to baseline noise include the detector and also the solvent delivery system. In contrast to most other chromatographic detectors the baseline of a UV-visible detector, in combination with the cell length, can be used to draw conclusions regarding the detectors sensitivity, without the need for injecting a standard. The baseline noise is reported as fractions of absorbance units. For the measurement the baseline is plotted for about 10 to 15 minutes after the instrument is equilibrated. The baseline noise is measured in sections of 0.5 to one minute and the results averaged. The noise is measured by using either a computer program or graphically. For graphical measurement pairs of parallel lines are drawn to form an envelope of all observed variations and the vertical distance between the lines is measured. Details can be found in ASTM method E-685-79. The difference in baseline noise with and without solvent flow gives information on the contribution of the solvent delivery system to the baseline noise.

Signal to Noise of Systems with Non UV-visible Detectors

For systems with electrochemical, fluorescence, refractive index and mass spectrometric detectors the signal to noise ratio is measured to judge a detector's sensitivity. One or more stable compounds should be selected that can be used to test the signal to noise over a long period of time under the same conditions.

Limit of Detection

This is established by injecting a standard diluted to such a concentration to produce a peak signal of about two times the noise. If the analyte can be easily detected over five injections, the system can be said to have a suitable limit of detection. Alternatively higher concentrations can be injected and the concentration is calculated that would correspond to two times the noise.

Limit of Quantification

This is established by injecting a standard diluted to such a concentration to produce a signal to noise of more than 20. The signal height and the baseline noise over a range representing a 20-fold peak width at half height are measured. The concentration is calculated that would correspond to 20 times the noise.

Linearity

Linearity is an important characteristics that influences quantitative results. The linearity not only depends on the detector characteristics but also on the type of compound and on the wavelength setting. Therefore linearity should be determined as part of method validation. A system's linearity is usually determined by the design of the detector. It is measured by injecting standards with at least four different amounts. The standards should span 50 to 150% of the expected sample concentration. Depending on the system's precision the standard may be injected once or several times at each concentrations. For multiple injections the responses (peak heights or areas) are averaged. Data can be evaluated mathematically, using a best linear fit regression analysis, and graphically. For graphical evaluation the averaged data are plotted versus the concentrations. Alternatively the averaged data are divided by their respective concentrations yielding the relative responses. A graph is plotted with the relative responses on the Y-axis and the corresponding concentrations on the X-axis on a log scale. The obtained line should be horizontal over the full linear range. At higher concentrations, there will typically be a negative deviation from linearity. Parallel horizontal lines are drawn in the graph corresponding to 95 percent and 105 percent of the horizontal line. The detector is linear up to the point where the plotted relative response line intersects the 95 percent line.

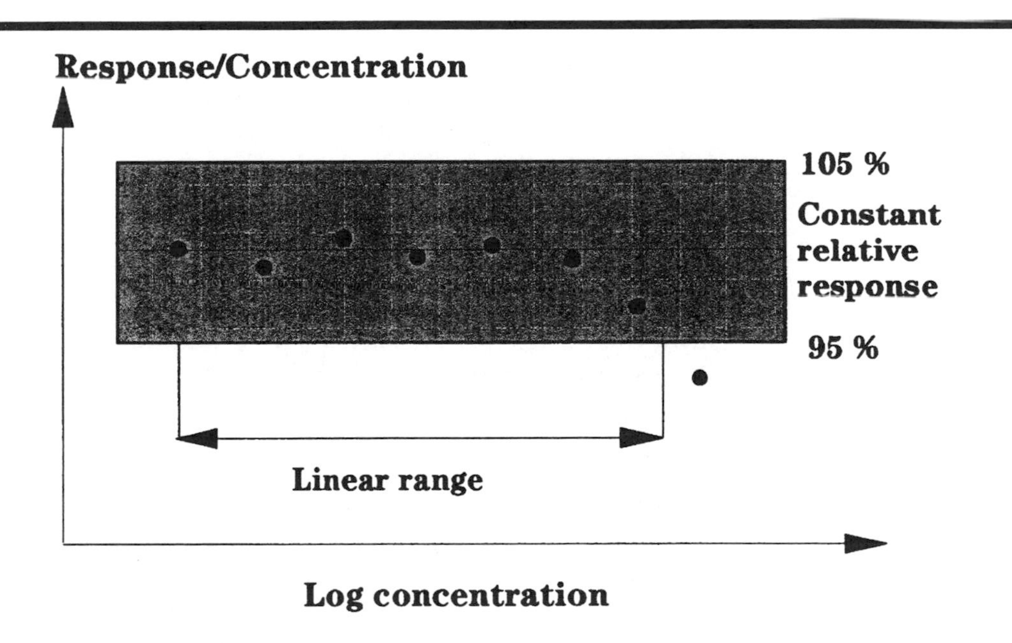

Figure D.2. Graphical evaluation of linearity

If for a particular analysis a variable volume injector is used the linearity should be also tested by injecting different sample volumes.

Wavelength Accuracy of UV-visible Detectors

The wavelength setting of a UV-visible detector can have an impact on important method performance characteristics such as selectivity, limits of detection and quantitation and linearity. The wavelength of a Variable Wavelength Detector is determined by the monochromator and the optical alignment of the detector. The wavelength may change through mechanical vibrations and should be tested and adjusted when installing the equipment and at regular intervals. Frequent wavelength accuracy check is specially required when the wavelength has been selected such that it is, for whatever reason, on the slope of the compound's spectrum. The accuracy can be tested by using a reference standard with known wavelength maxima. The wavelength maximum is measured with the instrument and the result is compared with the theoretical value. For detectors with scanning capabilities, the scan function can be used to find the maximum. For other detectors the wavelength is changed nanometer by nanometer around the expected wavelength maximum and the detectors response recorded for each wavelength. The standard may be delivered to the flow cell as a solution by pushing it through the cell with a syringe. Modern detectors employ a holmium oxide filter that has characteristic spectrum with a wavelength maximum at 361 nanometer. The filter can switched into the flow path at any time. The instrument then measures the absorbance maximum and compares the result automatically with the theoretical value.

Carry Over

Sample carry over from run to run can be a problem when the analyte amount changes significantly between runs. A carry over of 0.1 percent causes an error of 10 percent if the ratio of sample amounts is 100:1 between two injections. It is measured by injecting a sample followed by injection of mobile phase. In the chromatogram of the later the region in which the analyte is expected is examined.

Flow Rate Accuracy

This test may be performed as a diagnostic tool when the retention times drift in isocratic analysis. It can also be an important factor when methods are transferred between different instruments. For testing a volumetric flask is used and the time taken to fill the flask with mobile phase is measured.

Composition Accuracy

This test is recommended as a diagnostics tool when retention times drift in gradient analysis and gradient methods are transferred between instruments. The column should be replaced by an empty capillary to create a backpressure. Mobile phase A comprises a solvent (e.g., isopropanol or methanol) and mobile phase B the same solvent with 0.5 percent acetone. The detector wavelength is set to 267 nm. A step gradient is run from 0 to 100% B and detector response is recorded. The baseline increases in steps. The step heights are calculated in percentages based on the 0 percent and 100 percent B step. The differences between the measured step heights and the theoretical value represent the accuracy. Alternatively, a linear gradient is performed over the entire range. The detector output should be linear over the full range. The difference between the measured and the theoretical values at each percent B represents the gradient accuracy. This procedure can also be used to obtain information on the gradient linearity.

Pressure Test

The stability of the pressure, measured before the injection system, in isocratic operation and the pressure profile during gradient operation gives information on the leak tightness of the system. Modern HPLC pumps employ a pressure sensor with analog signal. The signal can be plotted before and during isocratic and gradient runs.

Capillary Electrophoresis (CE)

A complete CE system comprises a high voltage source, buffer containers and electrodes, a detector, a capillary housing and thermostat. Most commercially available systems include all these features in a single instrument. Some systems, usually constructed "in-house", consist of modules but retain the same features.

Migration Time Precision

Migration time precision is an important characteristic which influences the qualitative results of an analysis. The migration time precision is dependent upon the reproducibility of the electroosmotic flow which is chemically rather than instrumentally dependent. With this in mind appropriate capillary preconditioning and buffer replenishment conditions must be specified together with capillary temperature. A standard should be injected around ten times and the absolute and relative standard deviations calculated.

Peak Area Precision

Peak area precision reflects the injection reproducibility of the instrument. In CE since separations are achieved through differing velocities of analytes, the corrected area or spatial area of the peak should be calculated and used for precision calculations. This is readily obtained by dividing the reported peak area by the peak's migration time. A standard should be injected around ten times and the absolute and relative standard deviations calculated from the corrected area data.

Carry Over

Similarly to HPLC, it is important to assess the presence or extent of run to run carry over. This may be done by injecting a blank after injecting a sample.

Detector Baseline Noise

Similarly to HPLC, a UV detection system's baseline noise is an important characteristic which influences the limits of detection and quantitation. Baseline noise should be calculated using the analytical buffer with the voltage applied and with no voltage to assess the contribution from buffer flow and thermal effects. The noise is reported in fractions of absorption units. For the measurement the baseline is plotted for about 10 minutes after the initial baseline rise seen with most

CE separations. The noise is measured in sections of around one minute and the results averaged. Signal to noise may be determined in a similar manner to HPLC detectors.

Wavelength Accuracy

Wavelength accuracy may be determined using a built in Holmium filter or a well characterized analyte (see HPLC).

Detector Linearity

Similarly to HPLC linearity is an important characteristic dependent upon the detector, the analyte and other operating conditions. Linearity should be determined as part of a method validation. A detectors linearity may be tested using at least four standards of different concentrations covering 50 to 150% of the expected sample concentration. For analytical linearity the results obtained from the use of standard solutions should be compared to that from blank sample spiked to different amounts.

Injection Linearity

The injected amount in CE is easily altered, usually via software control. CE systems usually offer two different modes of injection, hydrodynamic and electrokinetic. For hydrodynamic injection the linearity of the injection parameters with corrected peak areas should be determined, preferably with a spiked blank sample or a real sample. With electrokinetic injection again injection linearity should be determined in a similar manner. However for an analysis, the inherent bias of electrokinetic loading and its sensitivity to sample matrix effects should be noted. Since the injection parameters between instruments from different manufacturers have not been harmonized, method transfer between instruments depends upon the absolute amount of sample injected being calculated and this amount translated to the injection parameters use in a different instrument.

Gas Chromatography

A complete GC system consists of an injection system, a flow or pressure controller, column oven and a detector. Similar to

HPLC important GC characteristics are the precision of peak areas, retention times, linearity and limit of detection. Quantitation and qualification in chromatography are achieved through comparison of a well characterized standard with the unknown sample. Therefore the precision between calibration runs of most performance characteristics such as injection volume, the gas flow rate and the column oven temperature (gradient) is much more important than the accuracy of these parameters. Accuracy of the column temperature and the flow rate are important if methods are transferred from one instrument to another one.

Precision of Retention Times and Peak Areas

Precision of retention times and peak areas are important characteristics that influence qualitative and quantitative results. They are influenced by the precision of the stability of the back-pressure regulator or the flow controller and by the column oven (gradient). A standard is injected about six times and the absolute and relative standard deviations of peak areas and retention times are calculated.

Signal to Noise of Detectors

The performance of gas chromatographic detectors is tested by measuring the signal to noise for specific compounds. One or more stable compounds should be selected that can be used to test the signal to noise over a long period of time under the same conditions.

Linearity

Linearity is an important characteristics that influences quantitative results. The linearity not only depends on the detector characteristics but also on the type of compounds. Therefore linearity should be determined as part of method validation. It is measured by injecting standards with at least four different amounts. The standards should span 50 to 150% of the expected sample amount.

Oven Temperature Accuracy

This test is important when it is critical to have retention time information closely matched between two or more gas chromatographs. The test can be used a diagnostic tool when the oven behavior is suspect. The temperature inside the oven is measured with a precision thermometer. The thermometer should be placed as close as possible to the oven temperature sensor.

References

1. US FDA, Compliance Policy Guide #7132a.07, *Computerized drug processing: Input/output checking,* October 1982. Available from FDA Freedom of Information Staff (HFI-35), 5600 Fishers Lane, Rockville, MD 20857.

2. US FDA, Compliance Policy Guide #7132a.08, *Computerized drug processing: Identification of 'persons' on batch production and control records,* December 1982. Available from FDA Freedom of Information Staff (HFI-35), 5600 Fishers Lane, Rockville, MD 20857.

3. US FDA, *Guide to inspection of computerized systems in drug processing* (The Blue Book), Washington, February 1983.

4. US FDA, *Software development activities: Reference materials and training aids for investigators,* Technical Report, US Government Printing Office, Washington DC, July 1987.

5. PMA, *Concepts and principles for the validation of computer systems in the pharmaceutical industry* (includes 33 papers), Proceedings CSVC seminar at Crystal City, Virginia, January 15–18, 1984.

6. PMA, *Concepts and principles for the validation of computer systems used in the manufacture & control of drug products* (includes 24 papers on contemporary computer validation subjects), Proceedings CSVC seminar II at Chicago, Illinois, April 20–23, 1986.

7. PMA's Computer System Validation Committee, Validation concepts for computer systems used in the manufacture of drug products, *Pharm. Technol.* 10 (5), May 1986, 24–34.

8. K.G. Chapman and J.R.Harris, PMA's Computer System Validation Committee, Computer system validation - Staying current: Introduction, *Pharm. Tech.*, May 1989, 60–68.

9. US FDA, Compliance Policy Guide #7132a.15, Chapter 32a, *Computerized drug processing: Source code for process control application software programs,* April 16, 1987. Available from FDA Freedom of Information Staff (HFI-35), 5600 Fishers Lane, Rockville, MD 20857.

10. E.J. Subak, Jr., PMA's Computer System Validation Committee, Computer system validation - Staying current: Software development testing strategies, *Pharm. Tech.,* Sept. 1989, 142–156.

11. J.S. Alford and F.L. Cline, PMA's Computer System Validation Committee, Computer System Validation - Staying current: Change control, *Pharm. Tech.,* Jan. 1990, 20–40.

12. J.S. Alford and F.L. Cline, PMA's Computer System Validation Committee, Computer System Validation - Staying current: Installation qualification, *Pharm. Tech.,* Sept. 1990, 88–104.

13. C.M. Schoenauer and R.J. Wherry., PMA's Computer System Validation Committee, Computer system validation - Staying current: Security in computerized systems, *Pharm. Tech.,* May 1993, 48–58.

14. UK Department of Health and Social Security (DHSS), *Good laboratory practice. The application of GLP principles to computer systems,* United Kingdom Compliance Programme, London, 1989.

15. C.J. Teagarden, A stepwise approach to software validation, *Pharm. Tech.,* Sept. 1989, 98–112.

16. T. Stiles, GLP and computerization, *Laboratory Practice,* 39 (5), 63–66 (1990).

17. K.G. Chapman, A History of validation in the United States, Part I, *Pharm. Tech.,* Oct. 1991, 82–96.

18. K.G. Chapman, A History of validation in the United States, Part II, Validation of computer related systems, *Pharm. Tech.,* Nov 1991, 54–70.

19. D.L. Deitz and C.J. Herald, Reconciling a software development methothology with the PMA validation life cycle, *Pharm. Tech.,* June 1992, 76–84.

20. US EPA, OIRM GALP Publication: *Good automated laboratory practices, EPA's recommendations for ensuring data integrity in automated laboratory operations with implementation guidance*, draft, Dec. 1990.

21. American National Standards Institute - Institute of Electrical and Electronic Engineers (ANSI/IEEE), *Software engineering standard 729-1983*, New York, John Wiley & Sons (1984).

22. American National Standards Institute - Institute of Electrical and Electronic Engineers (ANSI/IEEE), *Software Engineering Standard 830-1984*, New York, John Wiley & Sons, 11–13 (1988).

23. ISO 9000-3:1991. *Guidelines for the application of ISO 9001 to the development, supply and maintenance of software*, International Organization for Standardization, Case postale 56, CH-1211 Geneve 20, Switzerland (1991).

24. M. Dorfman and R. Thayer, *Standards, guidelines, and examples on system and software requirements engineering*, Washington, IEEE Computer Society Press, ISBN 0-8186-8922-6 (1990).

25. British Standards Institution, *British standard guide to specifying user requirements for computer based systems*, BS6719, 1986.

26. Canadian Standards Association, *Basic guidelines for the structure of documentation of system design information*, Canadian Standard Z242.15.4-1979, 1979.

27. National Bureau of Standards, *Guideline for life cycle validation, verification and testing of computer software*, FIPS Pub. 101, June 6, 1983.

28. European Space Agency, The software life cycle; the users requirements definition phase; and the software requirements definition phase, in *ESA Software Engineering Standards*, ESA PSS-05-0, Issue 1, Jan. 1987, 6–26.

29. National Aeronautics and Space Administration, Product specification documentation standard: Concept DID: SMAP-DID-P100, in *Product Specification, Documentation and Data*

Item Description (DID), Volume of the Information System Life Cycle and Documentation Standards, Release 4.3, Feb 2, 1989, 48–52.

30. S. Weinberg, R.M. Romoff and G. Stein, *Handbook of System Validation*, Weinberg, Spelton & Sax, Inc, Philadelphia, PA, 1993.

31. M.E. Double and M. McKendry, *Computer validation compliance*, Buffalo Grove, IL, Interpharm, 1993, ISBN 0-935184-48-1.

32. R. Chamberlain, *Computer systems validation for the pharmaceutical and medical device industries*, Libertyville, IL, Alaren Press, 1991, ISBN 0-9631489-0-7.

33. T. Stokes, R.C. Branning, K.G. Chapman, H. Hambloch and A. Trill, *Computer validation practices, common sense implementation*, Buffalo Grove, IL, Interpharm, ISBN 0-935184-5-4, 1994.

34. C. DeSain, *Documentation basics*, Buffalo Grove, IL, Interpharm, 1993, ISBN 0-943330-30-0.

35. R.F. Tetzlaff, GMP documentation requirements for automated systems: part III, FDA inspections of computerized laboratory systems, *Pharm. Tech.*, May 1992, 71–82.

36. A.J. Trill, Computerized systems and GMP - a UK perspective: part II, inspection findings, *Pharm. Tech.* March 1993, 49–62.

37. A.S. Clark, Computer systems validation: an investigator's view, *Pharm. Tech.*, Jan 1988, 60–66.

38. R. Black, EC & FDA legislation and validation requirements, paper presented at the *Course on Computer Validation*, organized by the International Society for pharmaceutical Engineering (ISPE), Stuttgart, Germany, Nov. 1993.

39. Commission of the European Communities, EC guide to good manufacturing practice for medicinal products in *The rules governing medicinal products in the European community*, Volume IV, Office for Official Publications for the European Communities, Luxembourg, 1992, ISBN 92-826-3180-X105.

40. M. Anisfeld, *International drug GMP's,* Buffalo Grove, IL, Interpharm, 1990, AT 23–24.

41. UK Pharmaceutical Industry Computer System Validation Forum (PICSVF), Good automated manufacturing practice in the pharmaceutical industry (GAMP), Conference and launch of Draft Guidelines: *Validation of automated systems in pharmaceutical manufacture.* Pharmaceutical Industry Supplier Guidance, Feb. 1994, Draft Version 1.0, in publication at LOGICA Industry Ltd. Units A & E, Business Park, Randalls Way, Leatherhead, Surrey KT 22 7TW, UK, Tel: + 44 71 637 9111, fax: +44 71 637 8229.

42. M. Rosser, Draft guidelines on good automated manufacturing practice: a conference report, *Pharm. Tech. Europe,* April 1994, 15–16.

43. Draft guidelines on good automated manufacturing practice: an extract, *Pharm. Tech. Europe,* April 1994, 18-21.

44. US EPA, Federal Insecticide, Fungicide and Rodenticide Act (FIFRA): *Good laboratory practice standards,* Fed. Reg 48, Nr. 230, 53946–53969, Nov. 29, 1983, effective: May 2, 1984.

45. US EPA, Toxic substance control act (TSCA): *Good laboratory practice standards,* 40 CFR Part 792, Fed. Reg 54, Nr. 158, 34034–34052, Aug. 17, 1989, effective: Sept. 18, 1989.

46. Organization of Economic Cooperation and Development, *Good laboratory practice in the testing of chemicals,* final report of the Group of Experts on Good Laboratory Practice, 1982, out of print.

47. European Community, *The harmonization of laws, regulations and administrative provisions to the application of the principles of good laboratory practice and the verification of their application for tests on chemical substances,* council directive 87/18/EEC, 1987.

48. European Community, *The inspection and verification of good laboratory practice,* council directive 88/320/EEC, 1988, adapted in 1990 (90/18/EEC).

49. Koseisho, *Good laboratory practice attachment: GLP inspection of computer systems,* Tokyo: Pharmaceutical Affairs Bureau, Ministry of Health and Welfare, 157–160 (1988).

50. OECD, Compliance of laboratory suppliers with GLP principles, *Series on principles of good laboratory practice and compliance monitoring*, number 5, GLP consensus document, environment monograph No. 49, Paris, 1992.

51. EC CPMP, *Good clinical practice for trials on medicinal products in the European Community*, Brussels, Commission of the European Communities, 1991, p. 23.

52. US FDA, Clinical Investigations, proposed establishment of regulations of sponsors and monitors, Federal Register 42, No. 187, Sept. 27, 1977, 49, 612–49.

53. US FDA, Protection of human subjects; informed consent; final rule, proposed establishment of regulations of sponsors and monitors, Federal Register 46, No. 17, January 27, 1981, 8, 942–8, 980.

54. US FDA, New drug and antibiotic regulations; final rule, Federal Register 50 (No. 36, February 22, 1985, 7, 452–7, 519.

55. US FDA, New drug, antibiotic, and biologic drug product regulations; final rule, Subpart D - Responsibilities of Sponsors and Investigators, Federal Register 52, No. 53, March 19, 1987, 8, 841–8, 844.

56. Code of Federal Regulations, Title 21, Part 50, Protection of human subjects, Washington DC, US Government Printing Office, April 1, 1991.

57. Good automated laboratory practices: where do instruments fit? *Analytical Consumer*, 3 (7), 10–11, (1993).

58. EN 45001:1989, *General criteria for the operation of testing laboratories*, Rue Brederode 2, B-1000 Brussels, CEN/CENELEC, The Joint European Standard Institution.

59. ISO/IEC Guide 25: *General requirements for the competence of calibration and testing laboratories*, 3rd edition 1990, International Organization for Standardization, Case postale 56, CH-1211 Geneve 20, Switzerland.

60. EURACHEM Guidance Document No. 1/WELAC Guidance Document No. WGD 2: *Accreditation for chemical laboratories: Guidance on the interpretation of the EN 45000 series of*

standards and ISO/IEC Guide 25,1993. Available from the EURACHEM Secretariat, PO Box 46, Teddington, Middlesex, TW11 ONH, UK, tel.: +44 81 943 7614, fax: +44 81 943 2767.

61. NAMAS, Accreditation standard, *General criteria for calibration and testing laboratories,* M10 of the NAMAS Executive National Physical Laboratory Teddington Middlesex TW11OLW, UK, 1989.

62. NAMAS, *A guide to managing the configuration of computer systems (hardware, software and firmware) used in the NAMAS accredited laboratories* - NIS 37 of the NAMAS Executive, National Physical Laboratory Teddington, Middlesex, TW11OLW, UK, Edition 1, 1993.

63. TickIT Office, *The TickIT guide to software quality management system construction and certification using EN 29001,* Issue 2, February 1992, ISBN 0-9519309-0-7. Available from the DISC TickIT Office, 2 Park Street, London W1A 2BS, UK, Tel: +44 71 602 8536, fax: +44 71 602 8912.

64. J. Guerra, Audits of computer systems in analytical laboratories, *Pharm. Tech.* 15 (9), 142–148 (1988).

65. F. Garfield, Quality assurance principles for analytical laboratories, Airlington, VA, AOAC International, ISBN 0-935584-46-3, 1991, p. 60.

66. US FDA, General principles of validation, Rockville, MD, Center for Drug Evaluation and Research (CDER), May 1987.

67. PMA Deonized Water Committee, Validation and control concepts for water treatment systems, *Pharm. Tech.* 9 (11), 50–56 (1985).

68. General Chapter <1225>, Validation of compendial methods, *United States Pharmacopeia XXII,* National Formulary, XVII, Rockville, MD, The United States Pharmacopeial Convention, Inc, 1990, 1710–1612.

69. D.C. Singer and R.P. Upton, *Guidelines for laboratory quality auditing,* New York, Marcel-Dekker, 1994, ISBN 0-8247-8784-6, p. 197.

70. A.J. Trill, Computer validation projects, problems, and solutions: an introduction, in *Computer Validation Practices,* Buffalo Grove, IL, Interpharm, ISBN 0-935184-5-4, 1994.

71. J. Agalloco, Validation of existing computer systems, *Pharm. Tech.,* Jan. 1987, 38–40.

72. N.R. Kuzel, Fundamentals of computer system validation and documentation in the pharmaceutical industry, *Pharm. Tech.,* Sept. 1985, 60–76.

73. H. Hambloch, Existing Computer Systems: A Practical approach to retrospective evaluation, in *Computer validation practices,* Buffalo Grove, IL, Interpharm, ISBN 0-935184-5-4, 1994.

74. P.D. Lepore, FDA's good laboratory practice regulations and computerized data acquisition systems, paper presented at the Sixth International LIMS Conference, Pittsburgh, PA, June 11, 1992.

75. ISO, *Guidelines for Quality Systems Auditing,* ISO 10011 Part 1.

76. L. Huber, *Good laboratory practice,* Waldbronn, Germany, Hewlett-Packard, publication number 12-5091-6259E, 1993, updated in 1994 (12-5963-2115E).

77. Hewlett-Packard, Good laboratory practice - Part 1: Vendor-validated computer-controlled HPLC systems, Waldbronn, Germany, Product Note, Publ. Number 12-5091-3748E (1993).

78. S. Christoph and F.M. Sakers, PMA's Computer System Validation Committee, Computer system validation - Staying current: Vendor-user relationship, *Pharm. Tech.* 48–52, Sept. 1993.

79. R. Sisk, Reengineering quality to meet customer needs, *Am. Lab.* 20, May 1994, 10–12.

80. K.G. Chapman, J.R. Harris, A.R. Bluhm, J.J. Errico, Source code availability and vendor user relationship, *Pharm. Tech.,* Dec. 1987, 24–35.

81. EN 45014:1989, *General criteria for suppliers' declaration of conformity,* Rue Brederode 2, B-1000 Brussels, CEN/CENELEC, The Joint European Standard Institution.

82. ISO/IEC Guide 22: *Information on manufacturer's declaration of conformity with standards or other technical specifications* 1982, International Organization for Standardization, Case postale 56, CH-1211 Geneve 20, Switzerland.

83. J.T. Abel, Computer system validation: questions to the audit, *Pharm. Eng.*, Sept./Oct. 1993, 50–59.

84. Hewlett-Packard, *Ordering Guide HPLC ChemStation,* Waldbronn, Germany, 1994.

85. W.B. Furman, T.P. Layloff and R.F. Tetzlaff, Validation of computerized liquid chromatographic systems, paper presented at the Workshop on Antibiotics and Drugs in Feeds, 106th AOAC Annual International Meeting and Exposition, August 30, 1992, Ohio, USA, published in *J. AOAC Intern.*, 77 (5), 1314–1318 (1994).

86. T. Layloff and P. Motise, Selection and validation of legal reference methods of analysis for pharmaceutical products in the United States, *Pharm. Tech.* Sept. 1992, 122–132.

87. E. Debesis, J.P. Boehlert, T.E. Givant, J.C. Sheridan, Submitting HPLC methods to the compendia and regulatory agencies, *Pharm. Tech.*, Sept. 1982, 9, 120–132.

88. H.B.S. Conacher, Validation of analytical methods for pesticide residues and confirmation of results, in *Pesticide Residues in Food,* 136–141, Lancaster, PA, Technomic Publishing Company, ISBN 87762-667-7, 1990.

89. L. Paul, USP perspectives on analytical methods validation, *Pharm. Tech.,* 15(3), 130–141, 1991.

90. F. Erni, W. Steuer, H. Bosshardt, Automation and validation of HPLC-systems, *Chromatographia,* 24, 210–207, 1987.

91. T.D. Wilson, Liquid chromatographic method validation for pharmaceutical products, *J. Pharm. & Biomed Anal.,* 8(5), 389–400, 1990.

92. Hewlett-Packard, Good laboratory practice - Part 2: Automated method validation and system suitability testing, Waldbronn, Germany, Publ. Number 12-5091-6082E (1993).

93. International Conference on Harmonisation (ICH) of Technical Requirements for the Registration of Pharmaceuticals for Human Use, *Validation of analytical procedures* (Draft consensus text), released for consultation Oct. 26, 1993, at Step 2 of the ICH Process.

94. G. Szepesi, M. Gazdag and K. Mihalyfi, Selection of HPLC methods in pharmaceutical analysis - III method validation, *J. Chromatogr.* 464, 265–278.

95. G.C. Hokanson, A life cycle approach to the validation of analytical methods during pharmaceutical product development, part II: The initial validation process, *Pharm. Tech.*, Sept. 1994, 118–130.

96. G.C. Hokanson, A life cycle approach to the validation of analytical methods during pharmaceutical product development, part II: Changes and the need for additional validation, *Pharm.Tech.*, Oct. 1994, 92–100.

97. US FDA, Validation of chromatographic methods, Rockville, MD, Center for Drug Evaluation and Research (CDER), Dec. 1993.

98. US FDA, Guidelines for submitting samples and analytical data for method validation, Rockville, MD, Center for Drugs and Biologics Department of Health and Human Services, Feb. 1987.

99. AOAC International, *AOAC Peer-verified methods program, manual on policies and procedures*, Airlington, VA, 1994.

100. V.P. Shah, et al., Analytical methods validation: Bioavailability, bioequivalence and pharmacokinetic studies, *Eur. J. Drug Metabolism and Pharmacokinetics,* 16 (4), 249–255, 1989 (1991).

101. H.T. Karnes, G. Shiu, and V.P. Shah, Validation of bioanalytical methods, *Pharmaceutical Research,* 8 (4), 421–426 (1991).

102. D.L. Massart, B.G.M. Vandeginste, S.N. Deming, Y. Michotte, L. Kaufman, in *Chemometrix—a textbook*, 1988, p. 114, Amsterdam, Elsevier, ISBN 0-444-4266-0.

103. L. Huber, *Applications of diode-array detection in HPLC*, Waldbronn, Germany, Hewlett-Packard, 1989, publ. number 12-5953-2330.

104. J.G.D. Marr, P. Horvath, B.J. Clark, A.F. Fell, Assessment of peak homogeneity in HPLC by computer-aided photodiode-array detection, *Anal. Proceed.*, 23, 254–257, 1986.

105. L. Huber and S. George, *Diode-array detection in high-performance liquid chromatography*, New York, Marcel Dekker, ISBN 0-8247-4, 1993.

106. J.C. Miller and J.N. Miller, *Statistics for analytical chemistry*, Chichester, Horwood, 1981.

107. Hewlett-Packard, *HPLC^{2D} ChemStation (DOS Series), Specifications*, Waldbronn, Germany, publ. number 12-5091-6524E, 1993.

108. N. Dyson, The validation of integrators and computers for chromatographic measurements, *International Laboratory*, June 1992, 38–46.

109. N. Dyson, The art of faking it, *Lab. News*, Oct 1991, 12–13.

110. J.K. Taylor, *Quality assurance of chemical measurements*, Chelsa, MI, Lewis Publisher, ISBN 0-87371-097-5, 1987, p. 195.

111. A.F. Hirsch, *Good Laboratory Practice Regulations*, New York, Marcel Dekker, 1989, 26–27.

112. R.T. Tetzlaff, Validation issues for new drug development: Part II, Systematic assessment strategies, *Pharm. Tech.* Oct. 1992, 84–94.

113. A.J. Trill, A regulatory perspective , in *Computer Validation Practices*, Buffalo Grove, IL, Interpharm, ISBN 0-935184-5-4, 1994.

114. M. Anisfeld, *International drug GMP's*, Buffalo Grove, IL, Interpharm, 1990, p. SA8–10.

115. N.R. Kuzel, Quality assurance auditing of computer systems, *Pharm. Tech.*, Feb. 1987, 34–42.

116. G.J. Grigonis and M.L. Wyrik, Computer System Validation: Auditing computer systems for quality, *Pharm. Tech. Europe*, Sept. 1994, 32–40.

117. S.P. Bruederle, CGMP's and the Pharmaceutical Laboratory, presented at the HP seminar *Road to Compliance*, March 1, 1994, in Chicago.

118. EN 45020:1993, General terms and their definitions concerning standardization and related activities, Rue Brederode 2, B-1000 Brussels CEN/CENELEC, The Joint European Standard Institution.

Name Index

Abel, J. T., 96, 241
Agalloco, J., 51, 235
Alford, J. S., 3, 102, 189, 190, 234
American Association of Pharmaceutical Scientists, 117
American National Standards Institute, 6, 35, 45, 235
American Society for Quality Contol, 185
Anisfeld, M., 14, 177, 237, 243
Association of Official Analytical Chemists (AOAC), 117, 185, 242

Biesel, Hans, 59
Black, R., 11, 236
Bluhm, A. R., 3, 71, 233
Boehlert, J. P., 117, 241
Bosshardt, H., 117, 241
Branning, R. C., 7, 191, 236
British Standards Institution, 6, 186, 235
Bruederle, S. P., 178, 244

Chamberlain, R., 6, 236
Chapman, K. G., 3, 5, 7, 38, 45, 71, 77, 191, 233, 234, 236, 240
Christoph, S., 71, 96, 240
Clark, A. S., 8, 57, 78, 236, 238
Clark, B. J., 120, 243
Cline, F. L., 3, 102, 189, 190, 234
Comité Européan de Normalisation/ Electrotechnical Standardization: The joint European Standards Institution (CEN/CENELEC), 186
Commission of the European Communities, 12, 13, 19, 74, 95, 152, 154, 236
Conacher, H. B. S., 117, 241

Debesis, E., 117, 241
Defense, U.S. Department of, 45
Deitz, D. L., 5, 189, 190, 192, 234
Deming, S. N., 119, 243
DeSain, C., 7, 236
Dorfman, M., 6, 186, 187, 235
Double, M. E., 6, 192, 236
Dyson, N., 146, 243

Environmental Protection Agency (EPA). *See Subject Index, page 248*
Erni, F., 117, 241
Errico, J. J., 3, 71, 233
EURACHEM, 26, 38, 115, 117, 129, 130, 132, 163, 238–239
European Community, 16–17, 188, 237
European Free Trade Association (EFTA), 188
European Space Agency, 6, 46, 235
European Organization for Testing and Certification (EOTC), 188
European Union, 16–17, 188

Fell, A. F., 120, 243
Food and Drug Administration (FDA). *See Subject Index, page 248*
Furman, W. B., 108, 123, 158, 159, 241

Garfield, F., 32, 239
Gazdag, M., 117, 122, 124, 242
George, S., 120, 243
Givant, T. E., 117, 241
Grionis, G. J., 177, 244
Guerra, J., 31, 177, 178, 239

Hambloch, H., 7, 51, 191, 236, 240
Harris, J. R., 3, 45, 71, 77, 233, 240
Health and Social Security, UK Department of, 3, 17, 234
Herald, C. J., 5, 189, 190, 192, 234
Hirsch, A. F., 155, 243
Hokanson, G. C., 117, 242
Horvath, P., 120, 243
Huber, L., 59, 120, 240, 243

Institute of Electrical and Electronic Engineers (IEEE), 6, 8, 35, 45, 235
International Conference on Harmonisation (ICH) of Technical Requirements for Registration of Pharmaceuticals for Human Use, 117, 242
International Organization for Standardization (ISO). *See Subject Index, page 249*

Japanese Ministry of Health and Welfare, 18
Johnson, Rick, 5

Karnes, H. T., 117, 119, 122, 242
Kaufman, L., 119, 243
Koseisho, 18, 237
Kuzel, N. R., 51, 177, 240, 244

Layloff, T., 108, 117, 123, 158, 159, 241
Lepore, P. D., 51, 177, 240

Margetts, Anthony, 5
Marr, J. G. D., 120, 243
Massart, D. L., 119, 243
McKendry, M., 6, 192, 236
Michotte, Y, 119, 243
Mihalyfi, K., 117, 122, 124, 242
Motise, P., 117, 241
Myers, Glenford, 11

National Aeronautics and Space Administration (NASA), 6, 45, 235
National Bureau of Standards, 6, 235
National Institute of Standards and Technology, 6, 190

Occupational Safety and Health Administration (OSHA), 190
Organization of Economic Cooperation and Development (OECD).*See Subject Index, page 250*

Paul, L., 117, 241
Pharmaceutical Manufacturers' Association (PMA). *See Subject Index, page 250*

Romoff, R. M., 6, 11, 236
Rosser, M., 16, 237

Sakers, F. M., 71, 96, 240
Schoenauer, C. M., 3, 234
Shah, V. P., 117, 119, 122, 124, 132, 242
Sheridan, J. C., 117, 241
Shewart, 134
Shiu, G., 117, 119, 122, 242
Singer, D. C., 41, 239
Sisk, R., 77, 95, 240
Software Requirements Working Group of the Software Engineering Standards Subcommittee on Software Engineering, 6
Stein, G., 6, 11, 236
Steuer, W., 117, 241
Stiles, T., 5, 234
Stokes, T., 7, 191, 194, 236
Subak, E. J., Jr., 3, 234
Szepesi, G., 117, 122, 124, 242

Taylor, J. K., 151, 162, 243
Teagarden, C. J., 4, 234
Tetzlaff, R. F., 7, 71, 78, 108, 123, 158, 159, 178, 236, 241
Tetzlaff, R. T., 177, 178, 243
Thayer, R., 6, 186, 187, 235
Trill, A., 7, 8, 50–51, 177, 178, 191, 236, 243

United Kingdom Department of Health Good Laboratory Program (GLP) Monitoring Unit, 17–18
United Kingdom Medicines Control Agency, 16
United Kingdom National Measurement Accreditation Service (NAMAS), 28, 115, 190, 239
United Kingdom Pharmaceutical Industry Computer System Validation Forum (PICSVF), 5, 15–16, 46, 74, 76, 85–86, 95, 108, 110, 191, 237
United States Pharmacopeia, 38, 117, 193
Upton, R. P., 41, 239

Vandeginste, B. G. M., 119, 243

Weinberg, S., 6, 11, 236
Western European Calibration Cooperation (WECC), 193
Western European Laboratory Accreditation Corporation (WELAC), 193–194
Wherry, R. J., 3, 234
Wilson, T. D., 117, 241
Wyrik, M. L., 177, 244

Subject Index

Acceptance criteria, 185
Acceptance tests, 5, 140
Accreditation, 185
Accuracy, 185
  of analytical method, 122
Algorithms, availability of details of, 94
Alpha testing, 65–66, 185
Analytical hardware validation, 85–91
Analytical instrumentation, performance verification of, 39–40
Analytical laboratory, validation processes in, 35–44
Analytical method
  range of, 123
  validation, 115–125
Analytical quality control, 185
Application of GLP Principles to Computer Systems, 3–4, 17–18
Application software, 185
  canned standard for, 72
  development and validation of complex, 200–203
  development and validation of simple, 197–199
  standard, 33
  user specific, 33, 72
ASTM method E-685-79, 224
Audit/inspection of computerized systems, 177–184
  audit questions, 182–184
  problems found during, 179–181
Audits, internal, 177
Audit tracking, 185
Audit trail for amended data, 160–161
Australian Code of Good Manufacturing Practice for Therapeutic Goods, 14–15
Automated liquid samplers, 174

Baseline noise of systems with UV-visible detectors, 223–224
Beta-testing, 66, 185
Black box testing, 34, 64, 79, 185–186
Boundary conditions, testing of, 65
Bug, 186

Calibration, 40–41, 128–129, 186
Canadian Standard, 6
Canned standard applications software, 72
Capillary electrophoresis (CE), 186, 227–229
  carry over, 228
  detector baseline noise, 228–229
  detector linearity, 229
  injection linearity, 229
  migration time precision, 228
  peak area precision, 228
  wavelength accuracy, 229
Carry over, 226, 228
Certified reference material (CRM), 133, 186
Change control, 69, 186
Checksum, 169, 186
Chromatographic computer system
  testing of, 139–150
    automated procedure for, 147–150
Chromatography. *See also* Gas chromatography
  detection limit, 123
  software for, 73
Combined standard and user contributed software, 80–83
Compliance, 187
Composition accuracy, 227
Computerized analytical systems
  retrospective evaluation and validation of existing, 210–216
  strategy for chronological selection and validation of, 217–219
Computerized system, 31–34, 187
  audit/inspection of, 177–184
    audit questions, 182–184
    problems found during, 179–181
  installation of, 101–105

logbook for, 105
operations of, 107–110
operator training for, 105, 107
revalidation and reverification of, 56–58
steps for routine use of, 110–113
Computer system, 2, 187
defining, 31–34
life cycle approach for validation of, 45–58
milestones in validation of, 4
security features built into, 167–169
Computer Systems Validation for the Pharmaceutical and Medical Device Industries, 6
Computer Validation Compliance, 6
Conformity, 187
Current Good Manufacturing Practice (cGMP) regulations, 12–16, 186

Data
back-up and recovery of, 167
entry of, 152–154
integrity of, 152, 163–167
security of, 163–167
validation of, 161–163
audit trail in, 160–161
data entry, 152–154
definition of, 151
raw data in, 154–160
Debugging, 187
Declaration of conformity, 88–89, 187
Declaration of System Validation, 187
Defective instruments, handling of, 135–137
Defect tracking and response system, 66–67
Detector baseline noise, 228–229
Detector linearity, 229
Detectors, 174
Diagnostics, error recording and reporting, 171–176, 187
Disaster recovery plan, 188
Documentation Basics, 7

EN 45001, 12, 188
EN 45000 series, 25–28
Environmental Protection Agency (EPA), 8, 46
Good Automated Laboratory Practices (GALP), 5, 20–24, 139, 154, 163, 189, 190, 235
Good Laboratory Practice (GLP) Standards, 1, 11, 16–19, 177, 178, 189, 237
interest in computer validation, 5
Office of Information Resources Management (OIRM), 5

Equipment, 188
baseline noise of systems with UV-visible detectors, 223–224
calibration, verification and validation of, 85–91
carry over, 226
composition accuracy, 227
flow rate accuracy, 227
limit of detection, 224
limit of quantitation, 224
linearity, 224–226
precision of retention times and peak areas, 223
pressure test, 227
signal to noise of systems with non-UV-visible detectors, 224
testing of selected, 221–231
wavelength accuracy of UV-visible detectors, 222
EURACHEM/WELAC Guidance on the Interpretation of EN 45000 Series and ISO/IEC Guide 25, 26–28, 129–130, 132, 163–164
Existing systems, evaluation of, 48, 50–54
External reference specifications (ERS), 61, 87, 188
External Standard (ESTD), 58

Federal Insecticide, Fungicide and Rodenticide Act (FIFRA), 16
Flow rate accuracy, 227
Food and Drug Administration (FDA)
Computerized Data Systems for Nonclinical Safety Assessment: Current Concepts and Quality Assurance, 3
Computerized Drug Processing; Identification of 'Persons' on Batch Production and Control Records, 2
Computerized Drug Processing; Input/Output Checking, 1
Computerized Drug Processing: Source Code for Process Control Application Software Programs, 3, 234
general principles of validation in, 35, 194, 239
guidelines for submitting samples and analytical data for method validation, 242
Guide to Inspection of Computerized Systems in Drug Processing (Blue Book), 2, 20, 71, 233
inspections of computerized laboratory systems, 7

Software Development Activities: Reference Materials, and Training Aids for Investigators, 2
on validation of chromatographic methods, 242
Functional testing, 64, 79, 188

Gas chromatography, 229–231
linearity, 230
oven temperature accuracy, 230–231
precision of retention times and peak areas, 230
signal to noise of detectors, 230
Good Analytical Practice (GAP), 9, 189
Good Automated Manufacturing Practice (GAMP), 189
Good Clinical Practice of Trials on Medicinal Products in the European Community, 12, 13, 19, 74, 95, 152, 154, 236
Good Computer Validation Practices, 7
Good Manufacturing Practice (GMP) regulations, 1, 11, 177, 178, 189
Good practice regulations, 93
Gray box testing, 65
Guideline for Life Cycle Validation, Verification, and Testing of Computer Software, 6
Guide to Good Manufacturing Practice (GMP) for Medicinal Products, 12, 13–14
Guide to Managing the Configuration of Computer Systems (Hardware, Software and Firmware) Used in NAMAS Accredited Laboratories, 28
Guide to Software Requirements Specifications, 6

Hardware, preventive maintenance of, 127
Hardware validation, comparison of software validation with, 45
Harmonization of Laws, Regulations and Administrative Provisions to the Application of the Principles of Good Laboratory Practice and the Verification of their Application for Tests on Chemical Substances, 16–17
Hewlett-Packard (HP), 8
Analytical Products Group (APG), 59
ChemStation, 31, 33, 96
Internal Defect Control System (DCS), 66
software contract service, 68–69
Software Tracking and Response System (STARS), 66, 192
Hewlett-Packard Life Cycle (HPLC) system
ChemStation, 79
detector flowcell, 85
diagnostic routines in, 173–174
examples for specifications and tests for a computerized, 141–145
testing precision of peak retention times and areas of, 204–209
High-performance liquid chromatography, 222–227

Injection linearity, 229
Inspection and Verification of Good Laboratory Practice, 16
Installation, 102, 104–105
preparing for, 101
Installation qualification (IQ), 41, 104
Instrument testing, 39
Integrated system testing, 143, 145
Intelligent error handling, 171–172
Interlaboratory test comparisons, 189
Internal audits, 177
Internal standards (ISTD), 58, 189
Inspection, 189
Installation qualification (IQ), 189
International Organization for Standardization (ISO), 6, 8, 46, 96, 189, 240
ISO 9000 series of standards, 11–12, 29, 189
ISO 9000-3, 6, 29, 46, 235
ISO 9001, 86, 96
ISO/IEC Guide 22, 88, 241
ISO/IEC Guide 25, 12, 25–28, 38–39, 40–41, 115, 135, 163, 186, 189–190, 194, 238

Laboratory accreditation, 190
Laboratory information management systems (LIMS), 9, 21, 32, 190
Life cycle (LC) approach, 2, 5
for new systems, 47–48
for validation of software and computer systems, 45–58
Limit of detection, 123, 224
Limit of quantification, 123, 224
Linearity, 224–226, 230
of analytical method, 122–123
Logbook, 105, 127, 172

Maintenance, 127–128
Manual data entry, 153–154
Master chromatogram, 148
Master data file, 148
Memoranda of understanding (MOUs), 17

Method validation, 42, 115–225
parameters for, 116–118
accuracy, 122
limit of detection and quantitation, 123
linearity, 122–123
precision, 120–122
range, 123
reproducibility, 120–122
ruggedness, 124–125
selectivity, 118–120
stability, 124
strategies for, 116
Microsoft®, 33
Microsoft® DOS, 72
Microsoft® MS-DOS®, 31, 33, 72
Microsoft Windows™, 31, 33
Migration time precision, 228
Modular functional testing, 141, 145
Modular testing, 108–109, 110

National standards, 190
New drug application (NDA), 9, 190
New systems, validation of, during development, 47–48

Obsolescence, 190
Office programs, validation of, used for laboratory applications, 54–56
Operation, 107–110
preparing for, 107
Operational qualification (OQ), 41, 107–108, 190
Operator training, 105, 107
Organization for Economic Cooperation and Development (OECD), 8
Compliance of Laboratory Suppliers with GLP Principles, 18
Good Laboratory Practice in the Testing of Chemicals, 16, 237
Good Laboratory Principles (GLP), 18, 71, 78, 79, 139, 238
Oven temperature accuracy, 230–231

Password, 190
Peak area precision, 228
Peak integration, verification of, 145–147
Performance qualification (PQ), 41, 110, 190
Performance verification (PV), 129–131, 190–191
of analytical instrumentation, 39–40
Pharmaceutical company, impact of validation on, 11
Pharmaceutical Inspection Convention (PIC), 12–13, 191
Guide to Good Manufacturing Practice for Pharmaceutical Products, 12
Pharmaceutical Manufacturers' Association (PMA), 8, 11, 46, 191
Computer System Validation Committee, 2, 3, 6, 41, 71, 76, 77, 96, 104, 163, 187, 233
Deionized Water Committee, 37, 50–51, 239
Pharmaceutical Technology Europe, 16, 237
Plug and play, 191
Precision, 120–122, 191
of retention times and peak areas, 223, 230
Pressure test, 227
Proficiency testing, 191
Prospective validation, 191
Pumping devices, 173–174

Qualification, 41
Quality assurance, 191
Quality control (QC) samples with QC charts, 132–135
Quality Management System (QMS), 29
Quality standards and guidelines, 25–29, 93, 177

Random access memory (RAM) test, 173
Range of analytical method, 123
Raw data, 154
audit trail for amended data, 160–161
defining and archiving chromatographic, 155–160
definition of, 154–155
R-charts, 134
Read only memory (ROM) test, 172, 173
Records, 191
Reference material, 191
Reference method, 122
Reference standard, 191
Registration, 192
Regulations, 11–24
Regulatory agencies, and accessibility of source code, 94
Relative recovery, 122
Reproducibility, 120–122
Retrospective validation, 48, 50–54, 192
Revalidation, 50, 56–58, 192
components in, 57
need for, 56–57
Reverification, 56–58
need for, 56–57

Risk assessment, 165
Ruggedness tests, 124, 192

Security of data, 163–167
Selectivity, 118–20
Signal to noise of detectors, 230
Signal to noise of systems with non-UV-visible detectors, 224
Software
  categories of, 72–74
  combined standard and user contributed, 80–83
  development standards for, 6
  revalidation and reverification of, 56–58
  standard application, 33
  subscription, 192
  testing, 64
  user specific application, 33, 72
  validation
    comparison of hardware validation with, 45
    definition of, 35
    steps for, 49–50
Software Engineering Standard 729-1983, 35, 235
Software Engineering Standard 830-1984, 6, 35, 235
Software life cycle, 8, 11, 59
  change control, 69
  design phase, 61–62
  documentation, 70
  implementation phase, 62–63
  operation and maintenance, 68–69
  release for production and installation, 67–68
  requirements analysis and definition phase, 59–67
  test phase, 64–67
Source code, 192
  availability of, 76–77
Specificity, 118. *See also* Selectivity
Standard application software, 33
Standard Operating Procedures (SOPs), 88, 192, 195–216
Stress testing, 65
Structural testing, 64, 192
Study director, 192
Study sponsor, 192
Suitability testing, 131
Support, role of vendor in providing, 79–80
System, 192–193
System life cycle, 193
System software, 32, 72
System stability, 124
System suitability testing, 43, 110, 131–132, 193

Test, 193
Test files, reuse of, 57–58
Testing, 38–39
  of chromatographic computer systems, 139–150
Test plan, 193
TickIT, 29
Toxic Substance Control Act (TSCA), 16
Traceability, 193

UNIX®, 33
Users
  responsibilities of, 71–83
  validation responsibilities of, 74–76
User interface prototyping, 193
User specific application software, 33, 72
User's site, verification activities at, 89–91
User testing, 79

Validation, 41–42, 194
  in analytical laboratory, 35–44
  of analytical methods, 115–125
  of data, 161–163
  definitions of, 35–38
  implementation phase, 62–63
    change control, 69
    documentation, 70
    operation and maintenance, 68–69
    release for production and installation, 67–68
    test phase, 64–67
  steps in the laboratory, 42–44
  at vendor's site, 59, 87–89
    design phase, 61–62
    requirements analysis and definition phase, 59–61
  versus verification, testing, calibration and qualification, 38–42
Validation of Automated Systems in Pharmaceutical Manufacture, 5, 74, 85–86
Vendor
  qualification of, 95–99
  responsibilities of, 71–83
  role of, in providing support, 79–80
  selecting, 93–95
  validation responsibilities of, 74–76
Vendor audits, 95

Vendor's site
validation efforts at, 59, 87–89
design phase, 61–62
requirements analysis and definition phase, 59–61
testing and specifications, 77–78
Verification, 39–40, 194
of peak integration, 145–147

Washington Conference on Analytical Methods Validation: Bioavailability Bioequivalence and Pharmacokinetic Studies, 117
Wavelength accuracy, 229
of UV-visible detectors, 222
White box testing, 34, 64, 194
WORM, 194
Worst case testing, 65, 194

X-charts, 134